Herbert Wendt was born in 1914 in Dusseldorf. The pro
of evolution fascinated him from his earliest student days out
he was prevented from following an academic career in zoology
by events in Germany in the 1930s and ultimately by the outbreak of the Second World War. Instead he became a writer
and today feels that this has in fact been the proper and more
satisfying path for him. Herbert Wendt is married with four
children. He also keeps eleven apes and many other animals.
His previous publications were *In Search of Adam*, *The Road
to Man*, *Out of Noah's Ark*, *It began in Babel*, *The Sex Life of
the Animals*, and *The Red, White and Black Continent*. His first
work, a novel called *Trennende Gitter*, was banned by the Nazis.

Herbert Wendt

Before the Deluge

Translated from the German by Richard
and Clara Winston

Paladin

Granada Publishing Ltd., 3 Upper James Street, London W1

First published in Great Britain by Victor Gollancz Ltd 1968
Published by *Paladin* 1970
Copyright © 1968 by Doubleday & Company, Inc.
This book was first published in German under the title
Ehe Die Sintflut Kam, © Gerhard Stalling Verlag, Oldenburg
und Hamburg
Made and printed in Great Britain by
Cox & Wyman Ltd., London, Reading and Fakenham

Contents

A Book with Seven Seals

The science of which this book will treat concerns itself with life on earth in distant ages. From fragments that have turned to stone it seeks to trace the evolution of this planet and its inhabitants. Its methods are the same as those used by archaeology and prehistory, but also are akin to those of criminology: industrious collecting, the finding and interpreting of clues, the amassing of evidence, and the application of imagination, deductive reasoning, and sheer inspiration. These methods make for its fascination; they sometimes also lead to fanciful, even fantastic conclusions which impose considerable strain on our credulity.

Goethe put it succinctly in *Faust*: 'My friend, the ages of the past are to us a book with seven seals.' But Goethe above all the men of his time understood the temptation mankind feels to break those seven seals and reach back to the remotest, most nebulous eras. That curiosity has often been a dangerous thing; the history of palaeontology is marked by constant conflict with scientific and ideological dogmas and authorities. 'Those few who have a little learning have ever been crucified and burned,' as it is put in *Faust*. As a result, the story of palaeontology has been a true-to-life novel of sensational discoveries, comical false trials, and revolutionary insights, but it has also been a succession of human entanglements, disputes, and tragedies. The human involvements will form the heart of our story. *Before the Deluge* is not a history of the antediluvian world, but a story of the men who explored that world, deciphering and interpreting its relics.

Scientists and philosophers have been interested in fossils since classical antiquity. But not until two centuries ago was it definitely established that fossils are the petrified remains of living creatures, often now extinct, which flourished in the earlier ages of the earth. For a long while scientists were simply baffled by the bizarre, primitive, and monstrous shapes that came to light out of the rubble of the remote past. The strange remnants preserved in stone seemed tokens of that mythic world-wide catastrophe which the Bible describes in the seventh chapter of Genesis: 'All the fountains of the great deep burst forth, and the windows of the heavens were opened.' Fossils seemed to confirm the legend of the Deluge; that is why to this day the creatures of remote ages are still often referred to as 'antediluvian'.

In the nineteenth century, when the theory of evolution began winning wide acceptance, palaeontologists were at last able to make sense out of the mythic world before the Flood, and reduce it to some system. But only in the present day, thanks to the precise methods of dating developed by physicists and chemists, has it been possible to arrive at a fairly reliable chronology of primordial times. The notion of the Deluge had to give way to the concept of evolution, and belief in sudden universal catastrophes was countervened by facts pointing to processes of gradual and almost imperceptible growth and decay. When, therefore, we use the word 'Deluge' here, we do so because it has become a symbol, as have the seven Days of the Creation, as have Paradise, Adam and Eve, the Tree of Knowledge.

How old is the earth? And how old is life? Ever since human beings have asked these questions, they have come up with the most astonishing variety of answers. Curiously enough, they have generally offered very specific figures, which have been refuted and smiled at by later generations. Five such figures may suggest the evolution of chronology.

In 1650 James Usher, the Irish archbishop, calculated the age of our earth from the evidences in the Old Testament, and concluded that it was created precisely at nine o'clock in the morning on 26 October of the year 4004 B.C. Around 1750 the French naturalist Count Buffon heated two large iron spheres red hot, and from the time they required to cool assigned to the earth a minimum age of 74,832 years. In 1862 Lord Kelvin, the British physicist, undertook similar experiments and arrived at a minimum of twenty million and a maximum of four hundred million years. In 1941 the American atomic physicist George Gamow raised this figure considerably. The firm crust of the earth, he wrote, had formed out of previously fiery-hot liquid rock 'approximately two billion years ago'. Nowadays, on the basis of the 'uranium clock', the rate of disintegration of radio-active substances in the earth's rocks, the age of the earth is estimated at somewhere between three and a half and five billion years.

What happened during these billions of years? The larger the figures with which geologists and palaeontologists reckon, the dimmer and more obscure do the remoter ages of the earth seem, and the larger the possibilities for serious errors. The animal and vegetable remains, which lie in the earth in the form of fossils, are often unnoticeable, shattered, fragile, and far more difficult to interpret than the remains of ancient human cultures. While archaeologists and prehistorians can base their conclusions almost entirely on human artifacts and traditions, palaeontologists often have only earth, stone, and tiny rudiments of organisms.

Scarcely any other science has had to work, from its very beginnings, with such sparse materials. Nevertheless, the palaeontologists have achieved amazing and often staggeringly comprehensive results. Despite the fragmentary objects of their research, and in the face of fierce disagreements within the fold and bitterest intellectual resistance from outside, they have so elucidated the history of the earth that today we can read it as we do human history passed down to us by the written word. The way they have accomplished this scientific miracle is the subject of this book.

Palaeontologists tend to be men obsessed, for whom digging and investigating is a passion. Often their work in palaeontology has been an avocation: officially they have been politicians, ministers, doctors, lawyers, engineers, artisans, miners. Some of the professionals have also been notable thinkers who have gone far beyond the boundaries of their field and made significant contributions to philosophy. On the other hand, palaeontological research has long exerted a magical attraction upon philosophers, writers, artists. Palaeontology is not only a branch of geological and biological science; it is a means of understanding and interpreting our world. Without it, we would be ill-equipped to place the tiny span which we call history within the vast stretches of universal time. In this 'science of the old' is united the sense of the past with the sense of the present.

HERBERT WENDT

Baden-Baden, Summer 1965

Book One

The Traces of the Deluge

Mountains can be annihilated, land can be conveyed from one place to another, peaks raised and lowered. The earth can open and close again, and even more things can occur which at first we would be inclined to regard as fables.

NICOLAUS STENO, *circa* 1670

GEOLOGIC TIME SCALE

ERA	PERIOD	EPOCH	SUCCESSION OF LIFE
CENOZOIC "RECENT LIFE"	QUATERNARY 0.1 MILLION YEARS	Recent / Pleistocene	
CENOZOIC "RECENT LIFE"	TERTIARY 62 MILLION YEARS	Pliocene / Miocene / Oligocene / Eocene / Paleocene	
MESOZOIC "MIDDLE LIFE"	CRETACEOUS 72 MILLION YEARS		
MESOZOIC "MIDDLE LIFE"	JURASSIC 46 MILLION YEARS		
MESOZOIC "MIDDLE LIFE"	TRIASSIC 49 MILLION YEARS		
PALEOZOIC "ANCIENT LIFE"	PERMIAN 50 MILLION YEARS		
PALEOZOIC "ANCIENT LIFE"	CARBONIFEROUS PENNSYLVANIAN 30 MILLION YEARS		
PALEOZOIC "ANCIENT LIFE"	CARBONIFEROUS MISSISSIPPIAN 35 MILLION YEARS		
PALEOZOIC "ANCIENT LIFE"	DEVONIAN 60 MILLION YEARS		
PALEOZOIC "ANCIENT LIFE"	SILURIAN 20 MILLION YEARS		
PALEOZOIC "ANCIENT LIFE"	ORDOVICIAN 75 MILLION YEARS		
PALEOZOIC "ANCIENT LIFE"	CAMBRIAN 100 MILLION YEARS		

PRECAMBRIAN ERAS	
PROTEROZOIC ERA	
ARCHEOZOIC ERA	

APPROXIMATE AGE OF THE EARTH MORE THAN 4 BILLION 550 MILLION YEARS

The Saint and the Sharks

Since 1965 a man of unusual character has come up for an unusual sort of trial. He lived three hundred years ago, and his case is being reviewed by the canon lawyers of the Vatican – who are not out to pass sentence upon him but to determine whether he is to be elevated to sainthood.

This man is the Danish physician and naturalist Niels Stensen. Under the name of Nicolaus Steno he occupies as prominent a place in ecclesiastical history as in natural history. His prospective canonization is truly astonishing, for Steno was a bold innovator in medicine and biology and has generally been regarded as the true founder of geology and palaeontology.

Our age of science has unfolded in what we may call a highly charged intellectual field, in which pure research into the facts of nature has always violently affected philosophy and religion. For centuries the development of science produced lively and often bitter conflicts between science and religion. The geologists and palaeontologists in particular touched dangerously on many religious and philosophical taboos. They brought about the collapse of whole cosmologies, and erected in their place new systems which seemed flatly contradictory to traditions and dogmas of the churches. For this reason many a scientist had to wage battle all through his life not only with scientific adversaries, but also with philosophical and theological authorities. Niels Stensen's transformation from naturalist to saint took place in a thoroughly unsaintly environment, in the brilliant city which in the age of the Renaissance was the indubitable centre of Occidental culture: Medicean Florence. The Medici, who dominated the city for three centuries, have earned their place in the history books not only by their financial activities, their wealth, their magnificence, their political intrigues, and the assassination of eighteen members of the family. They are above all remembered for their vigorous support of the arts and sciences. The most celebrated painters and sculptors of the age, among them Michelangelo, worked for the Medici. Galileo, the greatest physicist of the West, was their protégé. Important physicians and biologists, such as Francesco Redi and Marcello Malpighi, worked in Florence under the patronage of the Medici, studying the circulation of the blood, the nature of generation and fertilization, the functions of the human body and the development of the germ cell in the egg.

For those times, these were incredibly bold advances into unexplored scientific territory. Hence it was almost inevitable that Ferdinand II de Medici should have invited Niels Stensen to Florence. For Stensen appeared to be a brother in spirit to Galileo, Redi, and Malpighi. The Tuscan prince must have thought highly indeed of the Protestant anatomist from Copenhagen, for he presented him with a house as an inducement to stay.

Stensen was only twenty-eight years old when he arrived in Florence. Nevertheless, he was already regarded as a scientific genius. During his studies in Holland he had carried out painstaking dissections on a sheep's head and succeeded in finding the outlet of the parotid gland – the Ductus Stenonis, as it is still called in his honour. In Paris he discerned that the great philosopher Descartes had a totally false conception of the human brain, whereupon he sketched out a new and correct anatomy of the brain, thus founding the subdiscipline of encephalology. Above all, he made intensive studies of the structure of the heart and muscles. His observation that the heart is nothing but a muscle ranks among the greatest and boldest anatomical discoveries of his age.

A scientist who applied himself to such detailed examinations of the heart, brain, and other phases of human anatomy could easily, in those days, fall under suspicion of heresy. Stensen's colleague Malpighi had learned that to his sorrow. When he set about examining the processes of generation and fertilization, scandalized contemporaries destroyed his laboratory. Stensen, moreover, was known to be friendly with the philosopher Spinoza who was regarded in orthodox circles as a notorious atheist.

Ferdinand II de Medici did not permit such considerations to trouble him. He had held a protecting hand over Galileo, and he similarly shielded his anatomists and biologists.

The Secret of the 'Tongue Stones'

Stensen became the archduke's personal physician and tutor of his sons. He accompanied Ferdinand on various journeys which afforded him opportunities for collecting a great variety of marine life. The very first studies and drawings he published in Florence dealt with a delicate subject – sex. He described the organs of reproduction in fish and the ovaries of mammals. But under the protection of the Medici he remained, unlike Malpighi, unassailed. He was able to pursue his studies peacefully in whatever direction they led him.

In the meantime he had converted to Catholicism, and henceforth called himself Nicolaus Steno. One of the maxims he formulated at that time suggests that in spite of his epoch-making discoveries he was never a scientific rebel, but a modestly humble student of nature and at the same time a deeply believing Christian. The maxim went:

> What we see is beautiful,
> what we know more beautiful,
> what we cannot grasp, most beautiful.

A special kind of fossil present in some Florentine collections attracted Steno's attention and led him to his geological and palaeontological studies. Fossils at that time were viewed as more or less curious sports of nature. Only a few eccentrics, as we shall see, regarded them as the petrified remains of living things. The ecclesiastical as well as the scientific authorities frowned upon this theory.

Steno, however, had become anything but a conformist since his conversion to Catholicism. He examined the fossils carefully. The accepted name for them was *glossopetrae*, 'tongue stones', because of their tonguelike shape. Such stones are found quite frequently in a good many rock strata. There is little about them to suggest an organic form, and they were considered to be conclusive evidence that fossils were merely strange whims of nature.

Nevertheless a few Italian naturalists, among them the geologist Fabio Colonna and the physician-poet Girolamo Fracastoro, has in the past voiced their suspicion that *glossopetrae* might be petrified shark teeth. Steno used their conjectures as a working hypothesis, and ascertained that there were indeed many amazing resemblances between tongue stones and the teeth of sharks. At last in 1667, after he had been in Florence for a year and a half, good luck sent a stranded shark his way. He was able to make a close study of its teeth, and became convinced that the tongue stones were not natural curiosities but must have belonged to actual sharks.

But what kind of sharks? And how had the teeth become embedded in rock?

The fossil shark teeth, Steno realized, differed in some respects from the teeth of present-day sharks. They must therefore come from forms of sharks no longer extant. What was more, these great fish had been present in the interior of a continent, which meant there must once have been a sea there. Proceeding from this evidence, Steno sat down and in the course of two years of intensive work wrote what is probably the most brilliant book on geological and fossil phenomena to appear in the early days of palaeontological research.

From Geologist to Ascetic

The book bears the circumstantial Latin title *De solido intra solidum naturaliter contento dissertationis prodromus* ('Introduction to a dissertation on the solid substance naturally contained within solids'). The *prodromus* of the title suggested that the book was intended only as a 'forerunner' to a larger and more comprehensive work. But it contained such a wealth of questions, ideas, and problems that several generations of geologists and palaeontologists were able to draw upon it for sustenance.

Steno put forth a number of ingenious ideas on the origin of rocks and changes in the earth's crust. He drew a clear distinction between fossils of organic origin and those of mineral origin. He pointed out that normally one layer of earth is deposited on another so that the deepest stratum should also be the oldest. But he also drew attention to the presence, in some places, of steeply vertical strata, and concluded that rifts, quakes, and other later changes may have influenced the stratification.

Above all, however, Steno observed that there were considerable differences among the various strata, and that the fossils in lower layers often differed greatly from those in the higher layers.

What Steno had worked out was a useful crude sketch of the sciences of geology and palaeontology. Unfortunately, he never wrote the great work he had planned. Soon after his revelation of the true significance of 'tongue stones', Steno abandoned scientific work. He was consecrated priest, served subsequently as bishop of the German city of Münster in Westphalia, and in a number of other high ecclesiastical positions in Germany. In 1685, at the request of the Duke of Mecklenburg, he was transferred to Schwerin, to minister to the few Catholics of the duchy. He considered that he had been divinely summoned to this ungrateful task, and for the last year of his life lived ascetically and – according to contemporary accounts – like a saint, under the most wretched and poverty-stricken conditions in the capital of the tiny North German duchy. There he died in 1686 after a short illness. His scientific works were forgotten; it remained for Alexander von Humboldt, a century and a half later, to recognize the importance of Steno and the magnitude of his achievement.

The result was that Nicolaus Steno made no impression on his time as a scientist, and founded no school. All alone he had forged his way to a high peak in a new science, but what he had achieved went unregarded. Similarly the discoveries that had been made two hundred years before Steno were forgotten. They had come about at all only because one of the greatest geniuses the world has ever produced began to ponder.

'Sea Shells Far from the Sea'

In 1484 Duke Lodovico il Moro summoned to Milan a man who was then thirty-two years of age, an extraordinarily versatile artist and technician who had already distinguished himself in Florence as painter, sculptor, musician, poet, mathematician, physicist, and architect. In Milan this man Leonardo, from the village of Vinci near Empoli, was to design an equestrian statue of Duke Francesco Sforza, to help build the cathedral, and especially to serve as engineer for the Moor's canal-building projects.

The creator of the *Mona Lisa* and the *Last Supper* was also the foremost scientific thinker in Christian Europe, the first philosophical investigator who took up the threads where they had been left by the naturalists of classical antiquity. He studied ballistics and aeronautics, speculated on the nature of water and the nature of light; he propounded a number of laws of physics and constructed a flying machine. He dissected corpses, which at the time was an act of almost suicidal courage, and published a volume containing 235 anatomical drawings. He studied the stars, animals, plants, and stones. Thus it was only natural that the project of building the canals of Martesana and San Cristoforo should have led Leonardo da Vinci to analyse the geology of northern Italy.

In the course of the operations he noticed something that many engineers before him had seen, without being able to explain. The digging uncovered endless fossils which were clearly those of marine creatures. Leonardo undertook systematic excavations and found, as he put it, 'edible mussels, snails, oysters, scallops, shattered crabs, petrified fish, and countless other species'. But there was no sea in the vicinity of Milan. How had these remains of marine life found their way into the soil of Lombardy? How had they become petrified?

The Magical Shaping Power

As early as the sixth century B.C., ancient Greek scientists and philosophers had puzzled over the origin and nature of fossils. They had made the same observation as Leonardo: from quarries and deep in the earth in sections of country far from the sea, petrified remains of sea shells and other marine creatures had come to light. Empedocles, Pausanias, Herodotus, and other Greeks saw only one possible explanation: the regions in question must once have been covered by the waters of the ocean. Theophrastus of Lesbos and Xanthos of Sardes actually worked out a detailed theory of fossils. From the fact that petrified fish were found in the deepest fossil-bearing strata of the soil, Anaximander of Miletus drew the conclusion that fish must be the oldest ancestors of the existing animal world and therefore of men as well.

Leonardo was familiar with the classical writers, chiefly the Roman successors of the Greek natural philosophers. By and large they had set forth the same ideas as their Hellenic predecessors. Pliny, Strabo, and Seneca had offered theories of the changing extent of the sea and the effects of floods upon the general aspect of the earth. The poet Martial had written a few epigrams on fossil insects enclosed in amber; he had comprehended that amber had once been liquid resin in which insects had been caught. Ovid, too, had mentioned 'sea-born lands' in a poem, and spoken of 'sea shells far from the sea'; but since he had no inkling of how distant in time lay the origin of such shells, he rashly threw in the fossil of an 'age-old anchor'.

By the Middle Ages these Greco-Roman insights and theories were forgotten. Petrifactions were no longer recognized as the remains of once-living organisms, but sports of nature, whims of creation – products of a mysterious *vis plastica* or *virtus formativa*, a shaping power which as it were playfully modelled in stone all sorts of animal and plant forms. Some medieval thinkers even invoked astrology, maintaining that fossils were created by mysterious forces emanating from the stars.

To a mind like Leonardo's, such far-fetched theories were bound to seem preposterous. The Greek idea commended itself as far more scientific. But the curious doctrine of the *vis plastica* derived from two scientific authorities to whose doctrines the Christian West clung with dogmatic rigidity: Aristotle and his Arabic commentator Avicenna. Even in Leonardo's time it was not altogether safe to contest such dogmas.

The Origin of Fossils

'If you wished to maintain that the sea shells in these mountains were engendered by nature with the aid of the stars,' Leonardo noted cautiously in his notebook, 'how would you explain that these stars succeeded in creating such shells of differing sizes, differing ages, and differing species at the same spot?' He continued: 'Such a view is hard to impose upon minds with much capacity for thinking.'

Leonardo could see no other explanation but that the fossils were remains of genuine living creatures. The present-day continents, he concluded, had often been flooded by the sea in past ages. These marine basins became the home of all sorts of marine animals. After the floods receded, the former sea bottoms gradually hardened and turned to stone. The sea shells filled with mud which in the course of long ages likewise was transformed into stone. 'Thus,' Leonardo wrote, 'the sea shells remained preserved between two petrified substances, that is, the one surrounding them and the one inside them.'

His theory of the origin of fossilized fish was no less intelligent. The fish decayed, he reasoned; in their place there remained a hollow in the mud. Later floods brought in new and different types of mud which filled the hollows 'and preserved the exact form of the animals which were embedded in the older mud'.

If Leonardo had not been a rationalistic thinker, his remarkably modern-sounding interpretation of the fossils might easily have become entangled with a theory that had long been pitted against that of the *vis plastica*. The authority behind this was not the Greek naturalists but the Biblical story of the Creation. Like Leonardo, proponents of this theory held that the continental lands had once been flooded by the sea. Like him, they considered fossils the remains of once living marine animals. But they had only one flood in mind: the Biblical Deluge.

'Deluge – It Cannot Be'

The Deluge theory, which was later to dominate all of palaeontological research for generations, and to be responsible both for many advances in the field and for hampering progress in it, had actually been first suggested two hundred years before Leonardo's theory of fossils. Its first important advocate seems to have been the Italian naturalist Ristoro d'Arezzo. Ristoro had started out as a credulous follower of Aristotle and the theory of *vis plastica*. The rays of the sun and stars, he thought, could produce all sorts of impressions in stone; by their action certain rocks became 'figure stones', fossils.

In the year 1282, however, this remarkable man made a radical break with the old astrological view. He wrote a treatise on the structure of the world in which he pointed out that the Biblical Deluge had left clearly visible traces on the continents.

And in fact, as proof that all was once covered by water, near a high mountain peak we dug up many bones of fishes, as well as forms which we would prefer to call snails, but which other people might call sea shells. On one mountain we found coloured sand, mixed with large and many small rounded stones whose shape betrayed their provenance in water; a further indication that such mountains were overwhelmed by the Flood.

This theory, although it harmonized so well with the Christian doctrine of the Creation, was strangely ignored during the Middle Ages and the Renaissance. Four hundred years later, however, during the Age of Baroque, it was taken up with zeal by the majority of academic natural scientists and theologians in the West. Yet in principle it had already been refuted by Leonardo around 1500. Leonardo's critique of the Deluge theory is still another proof of his astounding gifts as observer and interpreter of nature. If later generations of scientists had built upon Leonardo's insights, palaeontology would have been spared many wrong turnings and blind alleys.

If you think that the Deluge carried these shells many hundreds of miles from the seas, I reply to you: That cannot be. For the Deluge was caused by torrential rainfall which the rivers naturally carried to the sea, along with the dead things that had been washed into them. The rainfall did not . . . draw the dead things from the shores of the sea to the mountains. . . . How would we explain, moreover, that in Lombardy rocks are found which are covered with corals and whole groups of oysters, that is, creatures that are accustomed to cling firmly to rocks? If the Deluge had carried sea shells three or four hundred miles from the sea, they would be thoroughly mixed in heaps with many other things. But instead we find the oysters all together, and also see the snails, the squid, and the other shellfish lying together in death as they lay in hordes while living, and we see the individual shellfish at just such distances from one another as we can observe on the sea-shores every day.

In other words, no sudden, disastrous flood could have cast the marine animals on to the continents. The fish and shellfish must long ago have lived where they were now found; what was today dry land had once been sea.

In 1517 Girolamo Fracastoro, the Italian physician and writer, arrived at a similar view independently of Leonardo. He saw the many fossils that came to light during the rebuilding of the walls of Verona and maintained that they must be the remains of creatures which had once lived on the spot. Fracastoro considered the theory of *vis plastica* absurd, the Deluge theory equally wrongheaded:

> They speak nonsense and folly who aver that fossils are remains of the Deluge in the Book of Genesis. Floods are transitory phenomena and in reality can only be caused by rivers. And how could such a flood leave the sea shells inside the stone? On the contrary, it would scatter the shells on the surface.

Fracastoro only touched on these questions. In Leonardo's notebooks, however, many pages are devoted to the key points of a planned 'Book on Petrified Sea Shells'. The draft of the most important chapter bears the heading: 'Refutation of those who maintain that sea shells were carried by the Deluge many days' journeys from the seas.' But these thoughts and theories were never published in Leonardo's time. He merely set them down in mirror writing for his own use.

When Leonardo's journals and notebooks, running to more than five thousand pages, were at last published, they could no longer influence the world of science. For palaeontologists had long ago come to the same conclusions as 'Faust's Italian brother'. It was Leonardo's tragedy that he had leaped ahead far in advance of his time, and for that very reason failed to influence the development of science.

The Restless Earth

A few years after the death of Leonardo da Vinci, a man named Georg Bauer, a native of Glauchau in Saxony, was tramping through the Erzgebirge (the Ore Mountains) studying stones and minerals. Bauer, who had Latinized his name into Georgius Agricola, was a physician by profession, like so many of the older naturalists. But his knowledge of geological matters was so impressive that in 1531 the ruler of Saxony, Electoral Prince Moritz, conferred a life pension upon him which enabled him to devote himself solely to mineralogical and metallurgical research. Agricola ultimately became the creator of German scientific mining procedures and the founder of the science of the earth in Germany.

Agricola was the first scientist to note that the earth is eternally subject to movement and change. Water, wind, ice, and fire dig valleys, raise mountains, and can cast up and remove again entire mountain ranges. Agricola recognized the work of erosion almost in the same terms as a modern geologist:

Small brooks, which at first only superficially wash the surface of the soil, cut deeper into the hard rock and in time can move great boulders. . . . The more the waters work their way into the depths, the higher the mountains rise on both sides. Rain and frost loosen small and large fragments and boulders from the mountains, which plunge down under their own weight. . . . The wind moves masses of sand, carries it rapidly along and heaps it up, so that under favourable circumstances it can coalesce and harden. In such wise the wind creates hills and mountains.

Agricola was also far in advance of his time in his opinions on the age of the earth. He was already thinking in very long periods, although he never cited specific figures; and like modern geologists he was convinced that the process of growth and decay on earth had in the past not been different from what they are today. 'But because human memory does not extend back into that remote past in which these mighty changes of the landscape began,' he wrote, 'most people think that there have not been such changes, although we can behold them going on still to this day.'

It was inevitable that Agricola should also turn his thoughts to 'figure stones'. Although he was unacquainted with the ideas of Leonardo and Fracastoro, he too surmised that these curious stones were not sports of nature, but the remains of marine creatures. Moreover, he coined a new name for the 'petrifactions', as such stones were then called. His word *fossil* – from the Latin *fossa*, a ditch – is still universally accepted.

In the literal sense 'fossils' are all the things which are dug out of the earth; the term was initially too inclusive. That became apparent ten years after Agricola's death, when the foremost descriptive naturalist of the day, the Zürich physician Konrad Gesner, published an illustrated octavo volume on petrifactions and minerals. Gesner included within the term 'fossils' not only the petrified remains of animals and plants, but also stone axes, lumps of ore, crystals, and even such archaeological finds as bracelets and jewelled rings.

Gesner took no more notice of Agricola's interpretation of fossils than of his geological theories. For to Gesner the world was immutable. On the other hand, Gesner did conceive a highly original classification of rocks and petrifactions. He arranged them neither alphabetically nor by substance, but by form. In his book *Fossils, Stones, and Gems*, he described linear, angular, circular, and elliptical stones and proposed curious and amusing names for the various types of fossils. Crinoids (sea lilies) were 'celestial forms', belemnites were 'thunderstones'. Gesner called fossil mussels 'sea stones', and petrified sea urchins 'Jewish stones'. He was a faithful follower of the doctrine of *vis plastica*, for which reason he took pains to point out that many fossils strikingly resembled 'the sun, the moon, and the stars'.

A Martyr of Palaeontology

In 1565, the year his book on fossils was published, Gesner died of the plague in Zürich. That same year a potter and glass-painter from Saintes, near the mouth of the Garonne, moved to Paris, where he began decorating the Tuileries with artificial grottoes full of strange terracotta animals and plants. His name was Bernard Palissy; he is known as the virtual father of French ceramics. In the course of his experiments in firing pottery, he once burned down his entire house with all its furniture. The rustic figurines which he finally succeeded in manufacturing were highly esteemed at the courts of Europe. Since Palissy was especially fond of depicting fish, crabs, mussels, and other marine creatures on his bowls, he searched the curiosity shops of Paris for specimens, and incidentally observed a wide variety of fossils. In this way the potter gradually became an eminent connoisseur of stones and petrifactions.

The amateur Palissy, who knew neither Latin nor Greek and who had never in his life read a work on natural science, hit on the same idea as Leonardo and Agricola. He concluded that the petrified shells of mussels and snails which turned up at every excavation in the Paris region must be the remains of actual animals, sea creatures that had once lived there. Hence Central France must at some period in the past have been a bay of the ocean.

In a book entitled *Discours admirables de la nature des eaux*, Palissy set forth this view. Possibly his ideas would have remained as unremarked as those of his predecessors if by ill luck he had not also been a sympathizer of the Huguenots. As long as he confined himself to pottery-making and experiments in enamelling, Catherine de Medici and other patrons could protect him. But when he published his book and began delivering scientific lectures on geology and petrified animals, no one could avert the charge of heresy.

Bernard Palissy became the first victim of persecution in the history of palaeontology. He was forced to recant, and with no consideration shown for his age – he was seventy-eight – was thrown into the Bastille as a Huguenot. There he died after two years, of malnutrition, consumption, and general debility. 'He lived like a good workman, he died like a saint,' Anatole France wrote of him three hundred years later. 'Can any man live better or die better?'

After what had happened to Palissy, the theory that the face of the earth had changed radically many times, and that fossils were the remains of living creatures, was heard less and less. Such ideas were still voiced and taught only at a few Italian universities, by scientifically trained physicians – who, however, prudently linked the doctrine with the legend of the Deluge, to avoid trouble with theological and secular authorities. The majority of scholars continued to regard fossils as curiosities and strongly repudiated any suggestion that they might be of organic origin.

In 1613 Martin Rinckhardt, a German pastor and hymn writer, composed a series of poems in praise of the young miners of Saxony. The cycle, entitled *The Christian Knight of Eisleben*, repeatedly mentions fossils. According to Rinckhardt, God has scratched all sorts of figures in stone, 'fish, men, and forms of beasts,' in order to show the men working in mines 'that He is everywhere with us'. This pretty if somewhat naïve theory was very widespread; it was not limited to miners, but shared by collectors, geologists, and physicians.

Dragons, Giants, and Unicorns

Even Stone Age man seems to have collected fossils. Finds in Peterfels, near Engen in Baden, in the vicinity of Vienna, and other places, indicate that primeval bands of hunters made ornaments out of petrified marine creatures. In an urn of ashes dating from the early Bronze Age, excavated near Bernburg in Thuringia, some fifty-five Tertiary sea shells were found. We have no idea, of course, what significance the people ascribed to these oddities which they made into ornaments or enclosed in tombs as grave goods.

The Chinese believed for thousands of years that the fossils of vertebrates were the remains of dragons. Such fossils were regularly collected and ground up for use as medicines. This belief in the curative powers of fossils spread from Asia halfway around the world. A few Chinese sages, however, suggested interpretations that sound strikingly modern. Thus, around A.D. 1200 the philosopher Shusi wrote: 'Once I saw shells in the rocks high in the mountains. I am quite certain that they were the shells of marine mussels. Thus those rocks must once have been ocean clay.'

Since earliest times, however, fossils far more impressive than those of fish and mussels have been unearthed: gigantic skulls and bones, mostly from ancestors of the elephant or the rhinoceros, as well as the bones of cave bears and other creatures of the Tertiary.

In ancient Greece such finds were considered the bones of dragons, giants, cyclopses, or centaurs. A skeleton ten ells long, dug up in the vicinity of Miletus, was identified by Pausanias as 'the bones of Homeric Ajax'. The philosopher Empedocles thought he recognized the skeleton of Homer's Polyphemus in the fossil bones of dwarf elephants that came to light in Sicily.

Giovanni Boccaccio, the author of the *Decameron*, described similar Sicilian finds in the fourteenth century. In the fourth book of his *Genealogy of the Gods*, he refers to these 'bones of the cyclops', and cites Empedocles as his authority for this view.

An elephant skull with its remarkable nasal opening, looking deceptively like two merged eye sockets, might easily cause an uninstructed observer to believe in the existence of gigantic one-eyed creatures. Othenio Abel, the Austrian palaeontologist, has commented: 'Sailors of Homeric days, blown off their course and finding such skulls in caves on the coast of Sicily, had never seen elephants. Hence they hit on the idea that the skulls must be those of huge beings with one great eye in the middle of the fore-head. Thus the myth of the one-eyed Cyclops probably origin-ated – as, indeed, most legends of giants are based on finds of large primordial mammals.'

As late as the second half of the seventeenth century fossil elephant skulls were still being disinterred from the soil of Sicily. A famous scientist, the Jesuit Athanasius Kircher, went to Sicily especially to see these skulls and examine the sites. In 1678 Kircher, whose fame rests chiefly on his studies of languages and classical antiquity, and on his invention of the burning mirror and the magic lantern, published a book entitled *Mundus sub-terraneus*. This account of the 'Underground World' became a milestone in the history of speleology, the exploration of caves and grottoes. Since caves are often rich in fossils, Kircher was led to interpreting the petrified remains of bones. But as if Steno had never written his *Prodromus*, Kircher decided all the fossils were the bones of giants.

Giants Bring Luck

Kircher provided his book with illustrations showing four types of 'giants' of differing sizes, including Homeric Polyphemus. Nevertheless, he was enough of a naturalist to observe that Poly-phemus had not been three hundred feet tall, as Boccaccio had asserted, but only thirty feet.

The aura of the legendary and uncanny that surrounded fossil bones was so strong that many fossils were exhibited in castles, cathedrals, and public buildings, or even suspended over portals.

The bones were thought to bring good luck; they were supposed to drive away evil, and often fetched high prices.

In 1443, when foundations were being dug for St Stephen's Cathedral in Vienna, the thighbone of a mammoth turned up. It was hung on one of the gates of the city, which henceforth bore the name 'Giant's Gate'. A mason chiselled the year 1443 into the fossil, and the motto of Emperor Frederick III: A.E.I.O.U. ('All earth is our [Austria's] underling'). The bone ended up in the Geological Institute of Vienna University.

In 1335, in a quarry near Klagenfurt, Austria, the fossil skull of an Ice Age woolly rhinoceros was found. It was considered a dragon's skull and exhibited as such in a curiosity shop. This skull, which today rests in the Klagenfurt Museum, was in 1590 used as a model by the sculptor Ulrich Vogelsang when he created the famous dragon monument that has become the great landmark of the city.

On January 11, 1613, near the castle of Chaumont in the Dauphiné, the skeleton of a dinotherium was discovered. This is a large, elephant-like beast of the Tertiary. The area around Chaumont had been known from very ancient times as the Champ des Géants, probably because many such fossils had been found there in the course of centuries. The find eventually came into the possession of a surgeon named Mazurier, who announced that the bones were those of the gigantic Cimbrian king Teutobochus. According to Mazurier, the skeleton had come from a grave inscribed with the name of Teutobochus, who in 102 B.C. was defeated in the battle of Aquae Sextiae by Marius the Gaul. For five years Mazurier travelled around exhibiting these bones for money. Meanwhile, the scholars at the Paris Academy wrangled – not over whether these were actually the bones of Teutobochus, but over the old problem of whether such bones were not, after all, sports of nature.

A good many of such 'giant's bones' mentioned in ancient accounts have been preserved to this day in collections, as was the case with the Chaumont dinotherium. Hence it has proved possible to identify them. In 1577, for example, some bones were found under the roots of an oak tree in the vicinity of Lucerne, Switzerland. The finders, assuming the bones were human, wanted to rebury them in the city cemetery. Thereupon a Basle doctor named Felix Platter asked permission to examine them. After a thorough study, Dr Platter declared that the bones were the remains of a giant who must have been about twenty feet tall. And since giants obviously did not qualify for Christian burial, the bones were placed on exhibition in the town hall of Lucerne. There they were seen two centuries later by Johann Friedrich Blumenbach, the German zoologist, who identified them as mammoth bones.

Almost every country has its national giant whose existence has been substantiated by some fossil bones. A mammoth bone hung above the entrance to the cathedral of Erfurt. Remains of fossil elephants were called the bones of giants in Belgium, while in Spain similar finds were revered as the bones of St Christopher. The mammoth, the great hairy elephant of the Ice Age, gave rise to tales about giants and beasts of fable in many lands.

One deposit, uncovered in 1645 at Hundssteig, near Krems in Austria, caused a great sensation. Swedish troops under General Torstensen were digging trenches there. What they found was described by the famous engraver and topographer, Matthäus Merian, in his *Theatrum Europaeum*:

About three or four fathoms deep they came upon a number of large, rotten, and partly broken bones, which were examined by learned and experienced persons and recognized as human bones. . . . The actual size of the predicated body is incredible; the head alone is as large as a round table, the arms as thick as the body of a man, and one tooth alone weighs five and one half pounds.

Merian showed two of the teeth in an engraving. From the picture, there is no doubt that they must have been mammoth molars. Some of the bones were sent to Sweden, some put on display in various Austrian collections and churches. A few of the teeth and fragments of bone were preserved at the imperial palace in Vienna, at the Church of St Nicholas in Passau, and in the oratorium of the Jesuit church in Krems. The last of the teeth was discovered in the Benedictine abbey of Kremsmünster in 1911, by someone who happened to know something about fossils. The astronomer of the abbey was innocently using it as a paperweight.

Swabians, who live in a land rich in fossils and are therefore much more familiar with them, did not leap to conclusions when they came upon mammoth teeth. In 1603 a mammoth tusk reputedly weighing five hundred pounds was found near Hall, in Swabia. The inhabitants of Hall fastened the strange object to St Michael's Church with iron bands, but they only placed a bit of verse beside it, the burden of which was: 'Now say, my friend, what I may be.'

The Pharmacist's Totem

A huge heap of mammoth bones found by a grenadier on the bank of the Neckar River, near Cannstatt, in April 1700, had an unusual fate. Duke Eberhard Ludwig of Württemberg ordered systematic digging in the area (probably the first such planned excavation in the history of palaeontology), and within six months seventy pieces of fossil ivory had been recovered. On orders from the duke a number of scholars assembled in Stuttgart to prepare an 'intelligible opinion' on 'whether these horns and bones are only a sport and work of nature, grown in the earth, or whether they were born in the womb of living animals, or whether all can be accounted human bones'.

The scholars who were to prepare this opinion were soon disputing fiercely. One would have it that the bones were the remains of Hannibal's elephants which had died in Germany after crossing the Alps. Another argued that they were bones from an old Roman sacrifice ground. A third held that the fossils were the 'effect and memento of the Deluge'. So bitter were the disagreements that the adversaries remained unreconciled to their dying days.

Nevertheless, the Swabian scholars were able to concur on a few points. They concluded that the Cannstatt bones were not of mineral origin, but belonged to the animal kingdom. For in 1703 Johann Samuel Kerl, a physician, burned some of the bone fragments. In the course of burning 'an evanescent urinous spirit along with stinking oil' was produced. This animal stench proved to Kerl's satisfaction that the remains were organic.

The controversial bones ended up in various collections, but the teeth were largely pounded into medicinal powder in the Stuttgart court pharmacy. This was because mammoth finds had revived the legend of the Biblical unicorn. Unicorn horn was obviously the best of all possible medicines. Unicorn pharmacies sprang up everywhere at this time. To this day the fabulous beast has remained the sign of the apothecaries' guild.

The legend of the unicorn probably stemmed originally from the sacred cow of the Orient. Then the rhinoceros of Asia and Africa became known in Europe. Pulverized rhinoceros horn is regarded as a panacea by a number of tribal groups, and similar medical properties were ascribed to the unicorn's horn. The long, corkscrew-like horn of the unicorn in traditional pictures of the beast was borrowed from the actual shape of a tusk belonging to the somewhat rare marine mammal called narwhal. Thus the unicorn of fable represents a cross of three animals: cow, rhinoceros, and narwhal.

When 'unicorn' was pulverized by apothecaries, or the horn cherished in the hoards of princes and dukes during the Middle Ages and early modern times, the material involved was usually narwhal tusk. The panacea was costly indeed. A duke of Saxony lightened his treasury by a hundred thousand talers for a single narwhal tusk. Emperor Charles V, who owed the Count of Bayreuth a large sum of money, discharged his debt with two narwhal tusks.

This tradition came to an end when the tusks of mammoths began being found along with the bones. Suddenly mammoth tusk was everywhere regarded as *unicornum verum*, true unicorn, while narwhal tusks were decried as *unicornum falsum*, sham and therefore worthless. Since unicorn powder was considered to heal all ills, disclose all poisons, and infallibly relieve impotence and sterility, countless mammoth tusks (and other fossils as well) were ground to powder. Such remains were avidly bought up by doctors, pharmacists, and quacks, and went by the name of *ebur fossile*, 'fossil ivory'.

Western man continued to believe in the existence of the unicorn as late as the seventeenth century. Consequently, there were various attempts to reconstruct the legendary beast on the bases of mammoth bones. In 1663 workers in a gypsum quarry near Quedlinburg came upon a great cache of mammoth bones and teeth. The Mayor of Magdeburg, Otto von Guericke, the famous physicist who invented the vacuum pump, attempted to reconstruct the skeleton of the unicorn from the fragments of bone.

The drawing he made of this reconstruction has been preserved, for the philosopher Leibniz printed it in his book *Protogaea*. Guericke's fantastic sketch of a skeleton bore not the slightest relationship to any real animal, but it was at any rate the first attempt to reconstruct a living animal from fossils.

An equally fantastic drawing of a fabulous beast was made by Captain Tabbert von Strahlenberg of the Swedish army while he was a Russian prisoner of war in Siberia in 1720. It was based on finds of mammoth remains in the taiga, and was again supposed to represent the unicorn. In the Siberian ice not only bones but well-preserved mammoth carcasses were discovered. Mammoth tusks were well known to the natives and Cossacks; called *mamontovakost*, they formed one of the most valuable articles of trade in northern Asia. But although countless ivory carvings were made of them, and although their resemblance to elephant tusks should have been immediately obvious to any scientifically trained person, European explorers continued until late into the eighteenth century to refer to them as 'horns of the Biblical unicorn', and drew pictures of them according to this pre-conception.

Those who did not regard the Siberian mammoth as a unicorn equated it with the behemoth mentioned in the fortieth chapter of the Book of Job. This was first reported by Nicolaus Cornelius Witzen, who as Dutch ambassador to Moscow in 1666 had gathered all available data on Siberian mammoths. In 1692, when Witzen was Mayor of Amsterdam, he wrote a book on his observations and inquiries. Commenting on the mammoth tusks, he wrote:

The tale is told here that these teeth are the horns of the beast behemoth, which the Russians call mammut or mammoth, of which mention is made in the Book of Job. The animal is said to live underground now and thus break off its horns, namely those teeth that are found. Its colour is supposed to be dark brown. It is said to be seldom seen, and when it does become visible the sight portends great misfortune.

Witzen himself correctly recognized the teeth as elephant tusks, and indicated his disbelief of the behemoth legend. Nevertheless, as late as 1714 the noted German-Russian physician and botanist Messerschmidt, who had found a complete mammoth carcass on the shore of the Indigirka River, had no compunctions about describing it as the behemoth of the Bible.

Big Bones in the Dragon Cave

What the mammoth was to the unicorn legend, the great Ice Age cave bear was to the dragon stories that play so large a part in European fable. The central European mountains are full of Dragon Hills, Dragon Caves, and Dragon Grottoes, with a variety of stories attached to them. Among the most celebrated are the Drachenfels in the Siebengebirge, where Siegfried is supposed to have slain the dragon of the Nibelungen saga; the Drachenhöhle near Mixnitz in Styria; the Drachenloch in the Tamina Valley of Switzerland. There are a large number of dragon caves in Swabia, the Alps, and the Carpathians. For all through these mountains huge bones and teeth have been found which a credulous populace could easily take as the remains of dragons.

A German physician named Patersonius Hayn, who had settled in Hungary, explored some of the dragon caves in the Little Carpathians ('White Mountains'). He found several curious skulls and a number of large teeth. Putting one of the skulls together (not quite correctly), he made a drawing of it, and in 1673 wrote a treatise for the Publications of the Academy of Naturalists in Halle entitled, 'Dragon Skulls in the Carpathians'. About the same time a German named Vette came upon similar bones in Transylvanian caves. A Halle naturalist named Vollgnad utilized his material for an article on 'Transylvanian Dragons', which were even supposed to have been able to fly.

The drawings accompanying these articles plainly show that both Hayn and Vette had discovered the bones of cave bears. The contents of other so-called dragon caves were of the same order. The cave of Mixnitz in Styria is to this day a rich source of cave-bear bones.

In scientific circles the question of the nature of fossils was discussed ever more heatedly. Were fossils mineralogical products, sports of nature, or were they the remains of fabulous monsters? Did fossil bones have special healing powers? Furthermore, was it possible that the supposed giants, dragons, unicorns, and so on, whose remains had apparently been brought to light, might still exist alive somewhere on earth?

All these questions, however, faded for a while after 1695, when the Deluge theory began its triumphant sweep through the world of scholarship. The great thing about fossils, then, was whether or not they confirmed the accounts in the Bible. There is a tendency among some writers nowadays to handle the evidence of archaeology in an exceedingly liberal manner, in order to prove that the Bible was 'right after all'. Similarly, in those days the facts of palaeontology were treated with equal looseness for the same reason.

The Might of the Waves

The book that was destined to revive the Deluge theory was actually a most sober scientific work, written by a learned member of the Royal Society in London – a man who taught physics at Gresham College and who had not the slightest intention of starting a conspiratorial society of sectarian fanatics. This professor of physics, John Woodward, had read Steno's great work on geology, and had found himself in disagreement with some of Steno's theories.

Steno had stressed that geological strata which lie one atop the other like the skins of an onion must necessarily have arisen at different times in the course of the earth's history. If these strata are not aligned in precisely horizontal positions, but are instead humped, folded or upthrust, Steno reasoned that this must have been due to local tectonic events, such as subsidences caused by 'underground leaching'. Woodward, however, did not think the strata were so very different. He refused to believe that they might have been formed at different times. Folds and other deviations from horizontal structure could be explained, he believed, as 'signs of terrible violence', such as an underground eruption of the sea. In short, all the stratification and changes in the earth, which Steno had called attention to, might be consequences of a great flood, a single world-wide disaster – the Biblical Deluge.

Monsters from an Unknown World

In his *Essay Toward a Natural History of the Earth*, Woodward dealt with the evidence for a universal deluge, as revealed by its effects. He was a very careful man and had studied the structure of the earth thoroughly before he ventured to publicize his views. He described his efforts to obtain as complete information as he possibly could about the 'entire mineral kingdom'. Whenever he heard of some remarkable natural cave or grotto, of a well being dug, or of pits where clay, marl, sand, gravel, chalk, coal, granite, marble, or metallic ores were being mined, he promptly set out for the spot. He carefully noted everything he found from the surface to the bottom of the excavation. In addition he drew up a questionnaire which he sent out to friends and acquaintances in places he could not manage to visit personally.

Woodward's notes reveal an almost pedantically exact investigator, certainly not a dilettante visionary. With equal deliberation he assembled all the evidence that challenged the notion of *vis plastica* and supported the theory of the organic origin of fossils. First he turned his attention to petrified marine animals. These marine shells looked exactly like their present-day relations on the shores of the seas, resembling the latter in shape, size, material, texture, sculpture, in the composition of the lamellae, the imprints of the muscle strands, and other respects. Moreover, the fossil shells showed one common characteristic of marine shellfish: smaller shellfish growing on the shells of larger ones. Thus, Woodward concluded, they must actually be the remains of animals that had once lived, and not merely stones.

The fossil bones of those immense creatures which were usually called giants, dragons, or unicorns might well be land animals, he said, but might equally well be creatures of the sea. The latter theory was held by the German philosopher Leibniz, who had heard about the discovery of mammoth bones and believed them to be 'bones of sea monsters from an unknown world which were transported by the might of the waves from the ocean to the land'. This 'might of the waves', which other thinkers repeatedly mentioned, must be identical with the Deluge, Woodward concluded.

As Woodward saw it, the great Flood had completely levelled out all the strata of the earth. When the waters receded, they did so with such violence that masses of earth and rock were raised high or gouged out. Thus mountains and valleys arose, and the layers of earth were displaced. The organisms trapped by the waters were forced into the stone; the might of the waves virtually petrified them. Such petrifactions, therefore, are the remains of animals that lived before the Deluge; but the ante-diluvian animals in no way differed from present animal life. Fossils that bear no resemblance to any living creatures must be the remains of animals that exist at present only in the depths of the oceans.

Woodward's essay, it is patent, is a curious compound of exact observation and absurd argument. In many passages the British physicist equalled Steno, Agricola, and other predecessors in objectivity. But his efforts to explain all geological formations as results of the Deluge sound forced and, compared with Steno's theories, totally unconvincing.

The Outpouring of Tartarus

Natural disasters such as the Deluge had become a subject for general speculation at the time, especially in England. In 1680 the appearance of a comet had aroused alarm, and the question was raised in a good many British publications whether the Biblical Deluge might not have been preceded or produced by such a comet. It may be that Woodward wrote his book under the influence of prophecies of universal doom.

Few thinkers in Europe during the Age of Enlightenment ventured to question the truth of the Flood and other such Biblical stories, as their predecessors had done. Rather, they made an effort not merely to believe the accounts in the Bible, but to prove them by empirical evidence and deductive reasoning. Leibniz was one of those philosophers who took special pains to analyse the legend of the Deluge and seek explanations for the event.

To a scientifically trained philosopher like Leibniz it was obvious that the Biblical story had one great flaw. Rain alone could never have produced a deluge capable of flooding the entire earth; if so, the atmosphere would have been so saturated with water as to be unbreathable. Leibniz therefore hit on the idea that there must be vast hollows filled with water in the interior of the earth. Mighty subsidences had squeezed this water out of the hollow spaces and 'the waters squeezed out of the cavities then flooded the highest mountains until they found new access to Tartarus, and after shattering the barriers of the hitherto locked interior of the earth, revealed anew what we today see as dry land'.

This theory was not so very different from Woodward's, which postulated the upsurge of an underground sea as the cause of the Deluge. Woodward's belief in the universal flood, and his attempt to prove it, were thus only in keeping with the spirit of his age. His great error lay in his doctrine that all fossils must be evidence for and victims of the Flood. For that postulate restricted palaeontology to a curious path, which was indeed to lead to great discoveries, but was soon to end in a blind alley.

An Old Sinner

The story begins with a piece of stone, ash-grey in colour, striated by eight black glossy vertebrae. Two Swiss students had found it on, of all places, Gallows Hill in the little university town of Altdorf, near Nürnberg. Gallows Hill bore its name for the simple reason that the town gallows stood there. Given this macabre site, it is understandable that the finders should have assumed the vertebrae were human bones.

But what kind of human bones? One of the students, Langhans by name, took it for granted that he had handled the mortal remains of a hanged man. Appalled, he threw the stone down the hill. But the other student, Johann Jakob Scheuchzer of Zürich, rushed after the stone and rescued a fragment containing two vertebrae. He added it to his growing collection of fossils.

This young man, Johann Scheuchzer, who was to become the most prominent and successful advocate of the Deluge theory, was in his student days still a faithful adherent of the doctrine of *vis plastica*. After completing his studies in medicine and natural history, he returned to his native city of Zürich, bringing back with him the fossils he had collected. In Zürich he worked as a teacher of mathematics and physician in an orphanage. He gathered around himself a group who shared an interest in natural studies, and delivered lectures about his fossils to the 'Collegium of the Well-disposed', as this band of sober citizens rather smugly called themselves. In keeping with the tendencies of the age, he identified most of the fossils as inorganic, mineral formations.

To the great surprise of the Collegium of the Well-disposed, however, Scheuchzer suddenly changed from a Saul to a Paul. He had read Woodward's essay, and was so impressed by it that he straightway translated the book into Latin. The translation was published in 1704, when Scheuchzer was just thirty-two years old. Henceforth, the Swiss naturalist gave full acceptance to the theory of the Flood, and he set about thoroughly re-arranging his collection of fossils and his ideas about them.

He began this operation by re-examining the pair of vertebrae from Gallows Hill in Altdorf. His friend Langhans had been right after all, he decided – they were human vertebrae. But Langhans had also been wrong – they were not the remains of a hanged man; they must be the bones of a man who had perished in the Biblical Deluge. Other finds that he had made in Franconia, in other parts of Switzerland, and on the shores of Lake Constance, fitted into this picture. Petrified fish, mussels, snails, and ammonites were not at all freaks of nature, they were marine animals that the great flood had long ago carried to the mountains. Scheuchzer became the standard-bearer of the Diluvians, who rapidly grew into a powerful scientific faction. Scheuchzer declared war on the rationalists among his contemporaries, who spoke of natural forces and laws of nature: 'I should rather cast my vote for a direct, divine, miraculous force than for a natural force.' He applied Woodward's conclusions to his native land: 'The period in which our Swiss mountains and all others arose was the time of the Flood.' The Deluge had, as he expressed it, converted the crust of the earth into 'a fluid jelly', of which undeniable evidence remained: 'the indisputable remains of the Flood locked into rock and soil: snails, sea shells, fish, and plants'.

The Advocate of Fishes

In order to put across the Deluge theory and thoroughly expose to ridicule the adherents of the old school, who held out for a *vis plastica* or even for the influence of the stars upon the rocks, Scheuchzer wrote a pamphlet entitled *Piscium querelae et vindiciae* – 'Complaints and Justifications of the Fishes'. In this polemic, published in 1708, a large pike of Lake Constance comes forward as spokesman of the fish and attempts to prove, in the finest Latin, that fossils are not 'mineral offspring of stone and marl', but remains of true living beings.

'We, the silent host of swimmers,' Scheuchzer's pike declares, 'herewith bring our complaints before the throne of Truth. We presume to demand what is ours and what has been wrongfully stolen from us by unsound philosophy or ill will, which support the tyranny of mineralogy.'

Fishes and other dwellers in water, the pike continues, suffered the effects of the Deluge through no fault of their own. They had to pay for human sins, and after the subsidence of the waters were left to die miserably on dry land. But men, instead of reorganizing their petrified remains for what they are, have degraded them, calling them lifeless matter, fragmentary 'figure stones'.

'We are fishes,' the pike proclaims in conclusion, 'not mere excavated fossils, not mere animal-like shapes sprung from the bowels of the earth, but a race borne along by the waves, living before the Flood and succumbing to it, victims of the madness of others.'

Along with his illustrations of fossil fishes, Scheuchzer published a drawing of the Altdorf vertebrae, labelling it the remains of a man who died in the Deluge, one of those guilty of the great disaster who had been its victim along with the innocent aquatic creatures. Unfortunately, a violent controversy sprang up over this specimen. One of Scheuchzer's former fellow students in Altdorf took issue with him. The man's name was Johann Jakob Baier; his importance to palaeontology derives largely from his obstinate feud with Scheuchzer. But that feud arose from a contrast that has occurred with great frequency in the history of science: the contrast between the fanatical adherent of a theory and an impartial, observant scientist.

Baier actually held the same views as Scheuchzer and his loquacious pike: that fossils must be the remains of organisms. And like his former fellow student, Baier had also collected fossils in Altdorf and the vicinity of Nürnberg. In a book entitled *Nürnberg Petrology* he described precisely and in detail typical Mesozoic fossils: ammonites, belemnites, and other extinct squids, mussels, and fishes. He assembled a number of proofs that 'ammon's horns' (as they were still popularly called) were 'not freaks of nature, but true shells and imprints of shelled animals'. In all these matters he was of the same mind as Scheuchzer.

Baier also did not doubt the 'universal great flood', for he too

45

was a child of his time. But he could not bring himself to link all fossils with the Deluge. And above all he mistrusted Scheuchzer's knowledge of anatomy. Baier was familiar with the Altdorf vertebrae. He had similar finds in his possession, called them *ichthyospondyli*, and was convinced that they were the vertebrae of fish.

Scheuchzer vehemently disagreed, and stuck to his assertion that the bones were human vertebrae. A heated correspondence ensued. Baier, for example, wrote to his famous friend and rival:

As for the vertebrae which you mention in your letter and which you show in a drawing in your *Piscium querelae et vindiciae*, I recognize immediately that they have the same form and constitution as my *Ichthyospondyli*. But I cannot grant that they are vertebrae from the backbone of a human being. I have not neglected to compare them with human vertebrae and I have made the observation that they differ considerably from these.

Baier then proceeded to expound the differences between piscine and human vertebrae, his tone implying that Scheuchzer had not the slightest fundamental knowledge of anatomy. He concluded sarcastically: 'I hope that your fish will forgive me for sending these vertebrae back to them, with your permission. If they look at them more closely, they will cease to claim them human and will perhaps discern what species of their own genus they actually belong to.'

Neither Baier nor Scheuchzer had any idea what species the bones actually belonged to. For the vertebrae were neither human nor piscine; they came in fact from an ichthyosaur found in the Jurassic black shale.

Baier's cavilings made no great impression on his contemporaries. Scheuchzer's views won acceptance. His triumph was due not only to his unswerving faith in the story of Creation as given in Genesis, but also to a new fossil site that had occupied him since 1706 and that was to provide him with his most famous discovery as well as his greatest mistake.

Fossil Herbaria

The western end of Switzerland's Lake Constance, which bears the name of Untersee ('Lower Lake'), throws two arms around a plateau-like mountain ridge. On the southern end, in the vicinity of the small village of Öhningen, great Tertiary strata have been deposited. In the Upper Miocene epoch, that is on the threshold from the middle to the more recent Tertiary, this region was covered with subtropical jungle. Forest lined extensive, shallow fresh-water lakes which left dense shoals of mud. Every palaeontologist nowadays knows that such a region is a treasure trove of fossils.

Around 1500, limestone began being quarried in Öhningen for building stone, floor slabs, and lime-slaking. The two quarries were owned by the Church, and passed successively from the monastery of Petershausen to the diocese of Constance and then to the canonry of Öhningen. The monks of Öhningen monastery provided the churches of the vicinity with floor paving cut from this limestone. Two of the greatest sculptors of the sixteenth and seventeenth centuries, the Dutchman Hans Morindt and the German Jörg Zahn, who worked for the Prince-bishop of Constance, were particularly partial to Öhningen limestone for their statuary.

The monks, and possibly the two sculptors as well, probably noticed long before the arrival of the first naturalists in Öhningen that the limestone slate of the quarries bore distinct imprints of fishes, sea shells, and other aquatic animals. Petrified frogs, snakes, birds, and other land animals must also have turned up in the stone from time to time. But the most striking feature of the Öhningen quarries was that some sections of them looked like fossil herbaria. The images of countless plant leaves were imprinted on the strata of limestone as if subjected to enormous pressure. Many were preserved to the most subtle details. The Swiss scientist Oswald Heer, one of the foremost palaeontologists of Europe, would later describe 475 different types of fossil plants from Öhningen alone. The majority of these plants were tropical or subtropical species.

Today we know why these quarries proved to be so rich in fossils. At one time, in the Upper Miocene, leaves and fruits dropped constantly into the shallow lake. They sank to the muddy bottom, along with the dead fish, giant salamanders, bullfrogs, snakes, and turtles which then inhabited the lake, and with the occasional land animals that drowned in it. The mud covered all these creatures quickly, forming an airtight layer over them; this mud subsequently hardened to stone which preserved the form of the plants and animals it had entombed down to their finest markings.

Until the beginning of the eighteenth century the monks of Öhningen did not know what to make of the delicate imprints, which they probably regarded only as natural curios. They did choose some of the most prettily marked slabs to use as new flooring in their monastery. But then two men from Switzerland came on a visit to them, and began quarrelling violently about the nature of these petrifactions. This dispute aroused the interest of the monks in their own fossils.

One of the two men, Dr Carl Nikolaus Lang, town councillor of Lucerne, was a fanatical adherent of the theory that petrifactions were created directly by the effect of 'seminal atmosphere' inside the earth. He zealously collected all sorts of minerals, crystals, stalactites, gems, and fossils and produced a learned book entitled *History of Figured Stones in Switzerland*. All these natural formations were due, he contended, to the influence of fine rod-shaped germs operating within the soil.

The other man was Scheuchzer. The monks took to him far more than to Lang, for he assured them that the apparent miracles of nature actually demonstrated the truth of the Bible. This also endeared him to the notables and gentlefolk of the vicinity, who showered presents on him: petrified oak, poplar, and laurel leaves, as well as fossil fish and other finds. As a result of Scheuchzer's work, fossils had acquired theological if not biological importance.

In Öhningen, Scheuchzer and Lang prudently avoided one another. Both men had recognized at once that the area was an ideal collecting ground for fossils. But Scheuchzer had won the monks' favour with his ideas about the Deluge. Consequently, any exciting new find went to him, assuring him of the lion's share of the fossils which came to light in Öhningen during the next twenty years. But it was not until 1715 that he obtained the stone slab that was to make him immortal – in a way that he certainly would not have relished. The slab contained, as we now know, the petrified forepart of a Tertiary giant salamander.

A Witness of the Flood

Scheuchzer at last had an articulated skeleton at his disposal. Triumphantly he proclaimed it to be the 'skeleton of a man drowned in the Deluge'. In his first publication on the subject he called it 'a very rare relic of that accursed race of men of the first world'. He promptly conferred a fine scientific name on the skeleton: *Homo diluvii testis*, 'the man who was witness of the Flood'. That man was, for Scheuchzer, also a principal witness in favour of his theory. The first copper engraving of the skeleton, which he had published in Breslau and London, appeared with a text that was intended to impress the fact upon his readers:

Alongside the incontestable witness of the divine word we have many other evidences for that terrible universal deluge: plants, fishes, four-footed beasts, vermin, sea shells, innumerable snails. But hitherto very few remains have been found of the men who were destroyed at that time. They floated lifeless upon the surface of the water and decayed; and from their bones, which are occasionally found here or there, it is not always possible to conclude that they are the bones of men.

But the Öhningen skeleton, he continued, was 'one of the surest relics of the Flood, indeed the most unequivocal'. It surpassed, Scheuchzer believed, 'all Roman, Greek, Egyptian, and other Oriental monuments in age and certainty'.

By using the data in the Bible, Scheuchzer had even attempted to calculate the exact age of this witness of the Deluge. He offered the conclusion that the 'man' had been drowned in the year 2306 B.C.

Scheuchzer's description of *Homo diluvii testis* created an enormous stir. Above all, it attracted geologists and collectors of fossils to the Öhningen quarries. And theologians, who had been prone to regard geological speculation with distrust, now hailed with satisfaction the evident harmony between the conclusions of science and the Christian religion. At the Sorbonne and other universities where heretical books had been burned and heretical writers of the calibre of Palissy banned, the academicians felt

49

vindicated after all. For had not Scheuchzer clearly proved that rationalistic theories of natural history were wrong and the Bible right?

From Scheuchzer's correspondence with an English naturalist, it would seem that in 1728 he found a second 'Deluge man' at Öhningen. But where this skeleton was placed or what it looked like has remained unknown.

Know, my dear, widely experienced sir, Scheuchzer wrote to his British colleague,

that some relics of the race of man who were drowned by the Flood have come down to us. Hitherto I possessed only two blackish glistening vertebrae of this race in my rather large collection. But then, by a fortunate turn of events, my museum was enriched with an impressive fragment found enclosed in the easily split stone of Öhningen. What can be perceived in this stone are not mere phantoms of the imagination. . . . Since I have become the owner of this memorable find, by unique divine Providence another such guest arrived at my door from the aforementioned quarry, one who is all the more welcome to me since he too possesses the virtue of extreme rarity and exceeds his former companion in size, age, and importance.

The first *Homo diluvii testis* was acquired by the Teyler Museum in Haarlem. At first no one doubted the skeleton's essential humanity. But when further skeletons of the same type were found in Öhningen, some murmured criticisms of Scheuchzer's identification began to be heard. A few investigators became convinced that the bones were those of a fossil catfish; others believed the 'witness' to have been a lizard. The mystery was finally solved a century later; in 1825 Georges Cuvier, the real founder of scientific palaeontology, travelled from Paris to Haarlem especially to examine the supposed 'Deluge man'. He immediately recognized the 'Witness of the flood' as a giant salamander of the Miocene epoch, a close relative of the still extant Japanese giant salamander.

Scheuchzer, however, believed to his dying day that he had won a total victory. At all the universities the Diluvians busied themselves filling bric-a-brac cabinets with fossils of all sorts. Expensive, magnificently illustrated volumes were published, full of fine pictures of 'antediluvian' creatures. Now and then one or another of his partisans turned up additional Deluge men and endeavoured, as a Diluvian expressed it, to recognize in the ancient deposits 'the countenance and other parts of the human body, as well as all sorts of other incredible things'.

During the last years of his life Scheuchzer published his *Physica sacra*, a four-volume 'Bible of Nature' which was intended to effect a union between the Holy Scriptures and natural history. He also devoted himself to collecting coins and to historical research, became a canon and Chief of Public Health in Zürich, and died in 1733 honoured by all. He outlived the theory of *vis plastica* by seven years; it had died a dramatic and ridiculous death in 1726.

Scandal

Almost all the books that deal with palaeontology or historical geology tell the story of poor old Professor Beringer of Würzburg – the tale of a highly respected man who fell victim to a silly students' prank and soon afterward died of grief.

Students, the story goes, manufactured all sorts of fossils, or rather made them out of fired clay, and buried them in places where Professor Beringer was in the habit of looking. Beringer reputedly found the forgeries, considered them genuine, and lovingly described them in a handsome folio volume. When, however, he finally dug up a 'fossil' bearing his own name, the farce was exposed. But it became a tragicomedy because Beringer could not survive the shock of having his lifework fall to pieces.

So much for the fable. It has been enriched by many ingenious details. According to one version, one of the forgers was the lover of Beringer's wife and who wished to expose the husband to shame in this unusual fashion. Then again, the suspicion has been voiced that the fraud was not the work of high-spirited students but of Jesuits who wanted to discredit a scientist inconvenient to them. Still other chroniclers aver that a man named Georg Ludwig Haeber was solely responsible, that he not only manufactured the forgeries, but also buried them and excavated them again; and that he even went so far as to write his absurd book in order to deceive Professor Beringer.

It ought to be apparent, from the above versions, that the case of Beringer is a first-class scientific mystery story – a highly unclear story full of obscurities and contradictions. What really happened in Würzburg between 1724 and 1726?

Deception of a Town

In reality the Beringer case, as the historians finally determined not very long ago, was a large-scale intrigue with prominent protagonists and even more prominent marginal personalities – a scientific scandal such as can hardly have happened before or since in a respectable university town. Among the chief actors was, beside Professor Johannes Bartolomaeus Adam Beringer, the noted historian Johann Georg von Eckhart, the founder of the science of prehistory in Germany. And one of the subordinate characters involuntarily involved in the affair was the great baroque architect Balthasar Neumann.

Let us examine the facts first. In 1724, Professor Beringer discovered at certain sites in the shell limestone formations around Würzburg a number of curious and often extremely realistic-looking figures: birds, lizards, frogs, fish, snails, crabs, worms, and flowers. Some of the frogs were coupling, one spider was in the centre of its web, another spider catching a fly. Gradually, more and more such 'figure stones' turned up: full moons and half moons, stars and comets, even Hebrew characters which spelled out the word Jehovah.

Neither students nor Jesuits led the professor to the sites and helped him with the digging. His guides and assistants were three peasant boys from the village of Eibelstadt on the Main: Nikolaus Hehn, Valentin Hehn, and Christian Zänger. The three boys came from poor families and were well paid by Beringer for their services. The oldest of the three was sixteen years old when the excavations started in 1724, the youngest twelve years old.

Simultaneously, the village lads seem to have carried on a thriving trade in the figure stones, supplying specimens to a considerable number of the citizens of Würzburg. Among their customers were the architect Balthasar Neumann and the Prince-bishop himself. Neither Neumann nor the Prince-bishop nor the other purchasers worried about the strange shape of the 'figure stones'; none of them seems to have wondered whether fraud might be involved. On the contrary, more and more eager buyers called on the three young peasants, so that the boys were gradually able to amass savings of three hundred imperial talers, at that time an extremely large sum. Evidently Johann Bartholomaeus Adam Beringer was not alone in his gullibility. Virtually half of Würzburg was being led by the nose.

Beringer was distinctly a man of parts. He taught natural history at Würzburg University and also served as dean of the medical faculty and personal physician to the Prince-bishop. Given his prestige and his multiple positions, it is highly probable that he was much envied. His general culture was not remarkable; his knowledge of Latin barely sufficed for him to write the book which later won him such notoriety. Contemporaries stressed his marked piety, but also his obstinancy and credulity.

Beringer had collected all sorts of freaks of nature long before the first fraudulent stones appeared. Mineral formations quite often exhibit remarkable resemblances to organic shapes. Curiously contoured lumps of metal, ore, marl, and minerals seem to imitate mosses or small, many-branched trees. A good many curiously shaped bits of sandstone or limestone resemble animal bodies. A 'basilisk', for example, which had been found at Number 7 Schönlatern Street, Vienna, in the year 1212, and which is mentioned in almost all the old books on fossils, was in reality nothing but a curiously shaped piece of sandstone. ... Beringer took pleasure in such sports of Creation, and rejoiced like a child over every newly discovered eccentric shape.

That is the point of the story. This pious, credulous, highly competent professional man never actually compared the 'fraudulent stones' with genuine fossils – regardless of everything that was said afterward. He regarded the stones simply and naïvely as miracles of nature. After all, the divine power had created gems as, in Beringer's words, 'mute but eloquent witnesses to God's perfection'. Why, then, should God in his omnipotence not also create stone plants, animals, and stars, and as it were sign them with his name in order to remind men of himself 'if they should forget their Creator'?

If we recall the clear and intelligent views of men such as Leonardo da Vinci, Fracastoro, Steno, Palissy, and Baier, and even the ideas of such Diluvians as Woodward and Scheuchzer, Beringer's faith in miracles, by comparison, seems distinctly simple-minded. But this means only that the Würzburg professor was not ahead of his time, as the above-mentioned thinkers were. On the contrary, he still had one foot in the past, in the Middle Ages.

Stones Bear Progeny

There is no reason to blame him for that. Many geologists of the time held even weirder views. They thought that minerals and metals were begotten by a 'stone seed'. Thomas Shirley, naturalist at the British court, wrote in 1762 that God poured into the bosom of the waters the seeds of all those things which later arose out of the water. The stone or rock seed contained in these waters, he wrote, penetrates into all bodies which come into the sphere of its activity and changes them. By its active fermentation it transforms every tiny particle of the matter it encounters into a perfect stone, corresponding to the idea inherent in it.

The most forthright advocate of this postmedieval version of the doctrine of *vis plastica* was Carl Nikolaus Lang of Lucerne, Scheuchzer's rival in Öhningen. The views of Shirley and Lang were repeatedly posed against those of the Diluvians whenever they spoke of the Deluge. At the University of Wittenberg a man named Schweigger proclaimed in a dissertation: 'I do not hesitate to assert that stones are created by their own seed and their own principles of generation. Stones are begotten as plants are begotten – on the principle of *like begets like*.'

Other naturalists attributed the origin of stones and fossils to the workings of 'petrifaction juices'. There were serious treatises on the digestive and reproductive organs of minerals; some stones were dubbed male, others female. A few years before the Beringer case, the book by one Valentini had been published in Frankfurt. Entitled *Collection of Objects and Curiosities*, it described among other things petrified sea urchins. If these sea urchins had thick spines, Valentini called them 'male stones'; thin-spined sea urchins were 'stones of the female sex'.

Stones copulated, could become pregnant, could bear young. Sometimes, in fact, stones brought their children into the world 'amid such travail that one could not sleep for the cries'. All such theses were set forth in dead earnest. There were so many theories, hypotheses, and conjectures that in the end geology capitulated to theology. Professor Etienne Declave plainly tendered his resignation from the tasks of science in 1635:

'There are so many contradictions among various views that it is impossible to resolve them all. If, therefore, we wish to treat things profoundly, as is necessary, we must step beyond the narrow bounds of natural science and take refuge in metaphysics, that is to say, in theology.'

Beringer did precisely that when he collected his freaks of nature, and when he was hoodwinked by the false fossils. He did not delve into the question of how these curious figures and signs had come to be. The principal thing for him was his faith that God must have been involved directly. In 1726, by which time he had seen some two thousand of these forged figure stones, Beringer published a report of the matter in the form of a princely volume, *Lithographiae Wirceburgensis*. Twenty-two copper plates depicted the two hundred finest of the 'script stones'. In these words, Beringer issued his appeal to the scientific world, asking its collaboration:

I prefer not to state my own opinion. I turn to specialists in order to be instructed in so puzzling a matter. . . . The learned petrologists will, it is to be hoped, cast light upon this question so obscure and unusual. Subsequently I shall add my own meagre illumination and will spare no pains to publish and exposit on whatever the soil of Würzburg may reveal in the future through the continued work of the excavators, and what I myself feel inwardly.

The Source of the Forgeries

Soon afterwards the bubble burst. It remains uncertain whether the exposure was precipitated by a figure stone inscribed with Beringer's name. Probably the ever more reckless commercial passions of the three Eibelstadt youths hastened the dénouement of the affair. But possibly the instigator of the forgeries himself went to the Prince-bishop after Beringer's book had been published and revealed the whole plot. If this last is the case, if the plotter hoped to drive the unfortunate Beringer out of Würzburg by disclosing that he had been taken in by silly forgeries, he found out to his dismay that the trick had boomeranged. For Beringer held on to his post, while the forger had to leave.

However the disclosure took place – on 13 April 1726 the three Eibelstadt boys found themselves in court charged with having falsified, buried, and sold the figure stones out of greed for profits. In the course of the protracted interrogations the Hehn brothers implicated Christian Zänger. They claimed that Zänger had misled them into burying the stones, digging them up again, and selling them. Zänger, in turn, cast the blame on two Würzburg professors; he declared that these two gentlemen had devised the plan in order to discredit Beringer 'because he is haughty and despises them'. The professors, said Zänger, had made the first forgeries and taught him how to produce the others.

One of the two accused professors was Johann Georg von Eckhart, a disciple and associate of Leibniz who in 1714 had come to Würzburg as historian and librarian of the palace and university. The other was a former assistant of Eckhart's named Ignaz Roderique, who had followed his teacher to Würzburg where he held a professorship in geography, algebra, and mathematical analysis. No more need be said about Roderique; he seems to have been little more than Eckhart's tool. But Eckhart's part in the whole affair is fascinating.

Eckhart had come from Hannover, where he had made a name for himself in history and prehistory. As historian of the royal house of Welfs (who in English are better known by the Italian form of the name, as Guelphs), he took up the task of his friend and mentor Leibniz after the latter's death. Emperor Charles VI elevated him to the nobility. Eckhart has won a place for himself in intellectual history as one of the first investigators of the ancient Teutons. He went about Lower Saxony examining many graves and barrows. Finding in the more recent graves chiefly bronze tools and weapons, in the older burial places implements of jade, nephrite, and other stones, he ventured to divide the early history of man into three great segments: the Stone Age, Bronze Age, and Iron Age. Thus Eckhart became the father of prehistoric studies in Germany.

Unfortunately, this excellent scholar seems to have had some rather grave defects of character. He was an envier, and could not brook playing second fiddle to anyone else. In addition, he led a rather disorderly life – to put it mildly. When he had run up such debts in Hannover that the city became too hot for him, he took to flight, and turned up in various monasteries in western Germany. Hitherto a confirmed and emphatic Protestant, Eckhart converted to Catholicism, apparently because he hoped for a post at some Catholic university.

This conversion was one of the principal reasons Beringer, pious Catholic that he was, could not endure Eckhart. Beringer regarded Eckhart as an opportunist; he thought – and probably rightly – that Eckhart was a man who trimmed his religious sails to every wind. Religious bias aside, Beringer also considered Eckhart 'an envious know-it-all'. Remarks Beringer made concerning Eckhart and his assistant Roderique have the salty tang of an age that still took joy in invective. He referred to his two colleagues as 'would-be sages with the cheek of certain foreigners', 'inexperienced novices in dealing with petrifactions', 'men who all too easily get beyond themselves', and so on.

Thus, the whole affair grew out of what was a fairly classical scholars' squabble. Eckhart knew that Beringer liked to go digging for pretty stones in the deposits of Würzburg shell limestone, and he knew that his adversary took pleasure in rare shapes and freaks of nature. He therefore conceived the plan of making a fool of Beringer by producing forgeries. Roderique carried out the work, and taught the technique to the peasant boy Christian Zänger, who proved a willing apprentice. Zänger in turn trained the Hehn brothers. Probably Roderique manufactured many of the fraudulent stones personally; Zänger probably went into fabrication chiefly in order to supply the demand of other collectors, such as the Prince-bishop, Balthasar Neumann, and other prominent citizens of Würzburg.

Eckhart and Roderique seem to have had no qualms about receiving the peasant boys in Eckhart's home after a successful coup. There, as Zänger testified in court, they 'all had a laugh and were delighted that Professor Beringer had been duped by these stones'. At that time Eckhart was a man of more than sixty and had just begun preparing the first scholarly edition of the great medieval German version of *Sohrab and Rustum*, the *Hildebrandlied*. It is really scarcely conceivable that a distinguished university professor in his sixties should sit down together with three callow village boys to gloat over a boyish prank he had played on a colleague. After Eckhart had – in Zänger's words – 'shown his exultation', he would give the boys a fat tip.

Eckhart was even audacious enough to take part in Beringer's excavations. When the first figure stones began being passed around in Würzburg, word of the discoveries reached other universities, where grave doubts were soon voiced about the genuineness of the finds. Eckhart thereupon volunteered to go along with Beringer as an eyewitness of the excavations, in order, so he explained, to refute all possible suspicion that anything might be amiss. Thus the trickster dug side by side with his victim for the fraudulent stones.

What the unsuspecting Beringer wrote about Eckhart sounds truly touching: 'He accompanied me to the fecund site where these marvellous things were found. . . . I allowed him to try the work, as entertaining as it was instructive, of digging these wondrous tokens of nature out of the soft soil with his own hands. . . . He could testify to having dug up more than one figure stone with his own hand.'

On these excursions sharp disputes frequently sprang up between Beringer and Eckhart. The historically trained Eckhart poked fun at Beringer's faith in miracles and pressed Beringer to consider whether the figure stones might not be amulets or talismans of the pagan Germans, which the old German magicians had perhaps hidden from Christian missionaries. Perhaps he was trying to lead Beringer by a roundabout route to the possibility that the figure stones might after all be worthless man-made things.

Beringer, however, merely became vexed with Eckhart's 'vain show of historical knowledge'. The very suggestion the figure stones might be products of Teutonic tribesmen made him cling more obstinately to his belief that they were 'divine wonders'. He seems to have had no very high opinion of Eckhart's Teutons.

Fossils and Counterfeits

After the trick had been exposed, Eckhart fell under the Prince-bishop's disfavour – primarily, no doubt, because the Prince-bishop himself had patronized the forgers. In addition, the

Prince-bishop was deeply offended with the two professors for having misused the name of Jehovah; he called Eckhart's conduct 'un-Christian' and cancelled the commission he had given him to write the history of the episcopal principality of Würzburg. He could no longer be certain, he commented, that Eckhart might not forge charters and other diocesan documents.

Eckhart retired to private life, and died four years later. Roderique left Würzburg and gave up his academic career for that of journalist. Balthasar Neumann wasted not a word on the falsifications, which had deeply disillusioned him. The Prince-bishop had the documents on the case sealed; they did not come to light again for two centuries.

Beringer alone continued to wage a lonely struggle. Legend has it that he realized how he had been deluded, and had his book bought up and destroyed, but that was not the case. In reality Beringer refused to draw the obvious conclusions from the scandal. He did have the first edition of his work bought up by a bookseller, but only in order to prepare a second edition to which he added a new preface. Undeterred, he clung to the opinion that most of the figure stones had been genuine and were in no way discredited by Eckhart's and Roderique's forgeries.

His reasoning was interesting. While admitting that 'envious persons had caused some forgeries to fall into my hands', in the next breath he declared: 'That does not make all the found figure stones false, any more than all old coins or all old works of art are false because counterfeiting and art forgeries exist.' He mustered all his eloquence to prove that it would have been impossible to forge so many image-bearing stones in so short a time. He contended that the sites of the finds had shown no traces of fresh digging. His thesis was that Eckhart had had a few figure stones forged in order to arouse the impression that all the finds were forgeries. He lashed out fiercely at his antagonist: 'And so this obscurantist and envier has in truly disgraceful fashion sought the dawn of his fame in calumny and deception.'

For fourteen years, until his death, the Würzburg professor strove in vain to rehabilitate his beloved figure stones. It was this tilting with windmills that gave rise to those legends which finally made Beringer totally ludicrous. And not only he, but also the entire scientific school which he – seemingly – represented.

Beringer has been painted as a fanatical believer in the *vis plastica*. The silliest views were ascribed to him, such as the possibility of spontaneous generation from stone, and the theory that his finds were 'the very first animals and plants become stone'. Yet Beringer never maintained anything of this sort. All he ever held out for was his faith in 'divine miracles'. But as a consequence of this fraud and Beringer's subsequent refusal to acknowledge it, the whole study of fossils for a while fell into disrepute. Was it not, after all, a dubious and shady business? How could anyone be sure that other petrifactions might not be the result of similar deceptions?

The Diluvians of the Scheuchzer school restored a good name to the science. They exposed the absurdities of all 'shaping forces', 'stone seeds', and 'stone births'. It was a good deal easier to believe in the Deluge than in Beringer's 'stone Creation'. The Beringer case, in fact, dealt the final blow to the doctrine of *vis plastica* and ushered in the victory of the Diluvian school.

Collectors, Patrons, Fossil Dealers

In the eighteenth century the science of fossils was furthered chiefly by the enthusiasm of amateur scholars. Rich men, and above all ducal collectors, were ready to pay a good deal for rare pieces to swell the array in their curiosity cabinets. The Jura Mountains in Swabia, South Germany, proved a treasure house for lovers of fossils. The sites at Bad Boll have been known since the end of the sixteenth century, and still yield interesting specimens today. Virtually the entire elite of 'petrifaction' specialists converged upon these mountains.

The well-known Tübingen chemist Johann Georg Gmelin obtained magnificent stone slabs containing crinoids (sea lilies) and belemnites from Bad Boll. From the same site, the royal house of Saxony bought 'the skeleton of an animal as large as a calf' for five hundred talers, and had it exhibited in the gallery of the Dresden Menagerie. It was a fossil crocodile. Students of Tübingen University collected so many petrifactions from the area that as one student of medicine named Strasskircher put it: 'fossils are ordinary household goods here'. Young Strasskircher actually lived by trading in fossils, earning his way through medical school in this way. Once he sold no less than twelve hundredweight of fossils to a Hamburg collector.

Even ladies of the vicinity fell to collecting. Probably the first female palaeontologist in the history of the sciences, the wife of a pastor named Essich, was extolled by her husband in the following tribute, written in 1714:

> With no little assiduity she turns over every rock, not sparing her tender hands, at once displaying the most accurate eye for the distinctions among fossils. Even her infant son, exceeding young and truly almost still in the cradle, is being guided to take long walks and collect figured stones, to which activity he is in any case drawn by inborn inclination.

In the Jurassic rocks of nearby Franconia, collectors worked with similar passion. Georg Wolfgang Knorr, engraver, botanist, and mineralogist, built up an impressive collection in Nürnberg. He described it in a four-volume work with the long-winded title *Collection of Curiosities of Nature and Antiquities of the Soil in Proof of a Universal Deluge.* This was the most thorough and significant palaeontological study of the times. Knorr died before it was finished, but the work was carried forward by his friend Johann Emmanuel Walch and concluded in 1775.

These early collectors were all, of course, Diluvians. But it is noteworthy that Walch, on the basis of his sound knowledge of fossils, held it possible that there might be changes and developments in organisms. Thus he was an early forerunner of the theory of evolution. In his edition of Knorr's *Collection of Curiosities* he pointed out 'that Nature imperceptibly passes from one genus to the other, placing between them a species which as it were represents the boundary between two genera and is therefore closely akin to both'.

One of the first great zoologists and explorers who hunted for fossils in faraway lands believed in the Deluge quite as a matter of course. He was Peter Simon Pallas, originally of Berlin. From 1769 to 1795 Pallas went on several extended expeditions for the Russian Tsar, travelling through all of northern Asia. He brought back a number of large fossil animals, including mammoth bones and the remains of an Ice Age woolly rhinoceros which he had dug out of the Siberian ice in 1773.

'The large bones which I found now scattered, now in skeletons, now heaped in hecacombs, so that I was able to observe their natural occurrence,' Pallas wrote in 1778, 'permanently convinced me of the reality of a catastrophic Flood that spread over our earth.' He postulated an underground source of fire under the Pacific Ocean which had raised vast masses of water and sent them rushing around the globe. 'This water pressed northward against the intervening mountain chains of Asia and Europe. ... Thus the remains of numerous large animals, that had been carried along in this general doom, were laid to rest.'

As is apparent from this quotation, the Diluvians were by now seeking more exact scientific explanations. Thus the curious situation arose during the second half of the eighteenth century that the adherents of the Deluge theory were far more progressive than their sceptical or atheistic opponents. In a mocking essay Voltaire contended that petrified fish and shellfish had nothing to do with the Deluge. 'It would be more natural to assume,' he wrote, 'that travellers, pilgrims, and crusaders took these fish and shellfish along for provisions and threw them away in the mountains because they had meanwhile spoiled, whereupon they subsequently petrified.' It is obvious that Voltaire offered this suggestion with tongue in cheek, solely to make fun of the Diluvian theologians.

When set against the Diluvians, not even the father of biological classification, Linnaeus, suggested vitally new directions to the science of palaeontology. In his *System of Nature* Linnaeus had arranged all the then known plants and animals, and provided them with scientific names. But when he also tried to do the same for fossils, his talents failed him. He devoted only a single page in his book to 'petrifactions'. Without comment, he listed Scheuchzer's 'witness of the Deluge', other finds from Öhningen, mammoth bones, fossil antlers of stags and elephant tusks. He placed all these in the Mineral Kingdom, withholding any mention of the fact that the scientific world had long ago accepted fossils as organic remains.

Later, to be sure, Linnaeus recognized that fossil mussels and snails turned up with singular frequency in limestone. 'Therefore I would surmise,' he argued, 'that these shelled animals

63

were not formed by the limestone begetting them, but rather that they themselves begot the limestone.' Here was the first time the hypothesis was explicitly advanced that stone might be formed from animal remains.

Petrifactions Exposed

The growing mania for collecting fossils and the intensive studies of these finds soon led to the first correct appraisals. One of these was made by an enthusiast named Balthasar Ehrhart, an assistant in Johann Georg Gmelin's pharmacy in Tübingen. As early as 1724 Ehrhart decided that belemnites, the 'thunderbolts' of popular literature, as well as the snail-shaped ammonites, could be neither snails nor mussels. He compared them rightly with the nautilus, a still extant cephalopod – in other words, a squid-like creature. This deduction was a masterly exercise of comparative anatomy – especially when we consider that prominent scientists long after Ehrhart continued to maintain that belemnites were crocodile teeth, or the prickles of gigantic sea urchins.

As early as 1715 an English apothecary and dealer in antiquities named Conyers described certain remains dredged out of the Thames as elephant bones. To be sure, he offered the theory that the bones must be those of a Roman elephant. In 1729 Franz Brückmann identified the 'giant's bones' dug up out of the loess of Lower Austria as the thighbones of elephants. Another sixty years were to pass before Johann Friedrich Blumenbach, the Göttingen professor of zoology, demonstrated that mammoths and other fossil elephants are not identical with present-day species of elephants.

Fossil rhinos also began to be identified. Blumenbach's predecessor at Göttingen, Professor Samuel Christian Hollmann, recognized as early as 1751 that 'dragon bones' from the Harz Mountains were actually rhinoceros bones. Twenty-two years later Pallas realized that the rhinoceros he had found in the Siberian ice differed from living members of the species, and Blumenbach corroborated this view. At the same time Johann Heinrich Merck, friend of Goethe, writer of the Storm and Stress period and official in the War Office of Hesse-Darmstadt, was taking so keen an interest in fossil pachyderms that he referred to himself as an 'elephant hunter and rhinoceros shooter'. In token of that, a Darmstadt zoologist, Johann Jakob Kaup, named a characteristic Interglacial rhinoceros after Merck.

Öhningen, where Scheuchzer had once discovered his *Homo diluvii testis*, continued to be a rich field for the fossil fancier. Its finds were put to good use by intelligent collectors and prosperous patrons. Outstanding collections were amassed by Privy Aulic Councillor Seyfried of Constance and the Prince-bishop Maximilian. The various dukes of Baden, monasteries and canonries, museums and private individuals in Germany and Switzerland eagerly bought up whatever petrified plants, insects, crabs, fishes, amphibia, reptiles, birds, and mammals came to light in the Öhningen quarries.

Ultimately, in the course of the nineteenth century, some of the world's most prominent palaeontologists came on pilgrimages to Öhningen to make a personal survey of the region. The world's foremost specialist on fossil fish as well as the discoverer of the Ice Age, Louis Agassiz, a Swiss scientist who later became an American, was among them. So also was Hermann von Meyer, founder of vertebrate palaeontology in Germany, and Oswald Heer, the Zürich botanist and entomologist. Even the British Museum took an interest in Öhningen and acquired, among other things, the almost complete skeleton of an Upper Miocene predator, famed in the literature as the 'Öhningen fox or civet'.

As is the case with all celebrated and productive sites, finds from Öhningen were especially prized. Extraordinarily large sums were paid for more 'Deluge men', although no one took them for that any longer. Equally good prices could be had for fossil bullfrogs, tortoises, snakes, and particularly fine specimens of tench, pike, eels, and other biggish fish.

When a site has been exploited for several centuries on end, it is to be expected that the same species of animals and plants will crop up again and again. Finds were repeated and prices consequently dropped. When something exceptional was unearthed, it was apt to be smuggled out rather than be duly delivered to the quarry owners. The result was that Öhningen soon produced its own forgery scandals – not just once, but several times.

The Fossil Workshop on Lake Constance

On orders from Prince-bishop Maximilian of Constance, a canon of Öhningen named Peter Pfeiffer began offering the quarry workers higher and higher finder's fees for rare and well-preserved fossils. Other collectors similarly bid up the prices. Perhaps the lessons of the Beringer case had been taken to heart and warnings were issued not to deliver any forgeries to the Prince-bishop. Then again, it is possible that no one in Öhningen had ever heard of the Würzburg events and the idea of setting up a fossil workshop arose independently, out of the circumstances.

In any case, the lay brothers apparently joined with the quarry workers to devise the best methods for preparing sham fossils. The workers simply wanted to improve their wages with bonuses for fossil finds. The lay brothers probably mingled a certain degree of malice with the profit motive: they enjoyed leading the seemingly learned and knowledgeable buyers, dealers, and collectors around by the nose.

The first Öhningen forgeries were probably produced around 1775. At first the forgers were conscientious enough not to construct artificial fossils. Rather they used fragments of fishes and other common fossils of low market value, and artfully assembled such fragments into good looking new skeletons of whole animals. They made similar use of plant fossils. The Öhningen quarries yielded so many imprints of poplar leaves that collectors were no longer interested in them. But if a workman carved away a bit at these imprints, he could produce unusual leaf shapes that were instantly saleable.

This puzzle game produced a number of new species and genera. The Prince-bishop's scientific advisers were no sharper at recognizing the forgeries than were other collectors. From Meersburg, Constance, Petershausen, Rheinau, and other cities and towns in the vicinity of Lake Constance, delightful showers of talers rained on Öhningen. The Meersburg Natural History Collection, the collection of Hofrat von Seyfried, and the curiosity cabinets of dukes and commoners who bought their fossils from Öhningen sources, soon held large selections of monstrous and odd animal shapes and foliage prints from the Öhningen workshop.

Success emboldened the forgers. They procured the bones of ordinary horses and pigs, arranged them suitably and carefully embedded them in the moist, relatively soft limestone. Ultimately they even concocted certain weathering agents and cements to make their products look as original as possible.

Decades passed before the forgeries were gradually exposed. The secularization of 1805 and 1806 transferred ownership of the quarries partly to the local peasants, partly to the State. The Meersburg Natural History Collection was broken up, as was the Seyfried collection; the fossils from these collections were sent to the Karlsruhe Museum of Natural History, where they remain to this day.

Karlsruhe scientists received the palaeontological materials from Öhningen with mixed feelings. On the one hand they were proud of these fossils, which had become world famous. On the other hand, in the very year 1805, when the secularization began, a teacher from Constance had published an exceedingly disturbing report on the Öhningen quarries and fossils. This report contained not only the first precise description of the geological conditions of Öhningen, but also the first references to forgeries.

Josef Maximilian Karg, a town councillor and teacher of natural history in Constance, was the author. Karg, evidently an excellent geologist, had familiarized himself with the operations of the Öhningen fossil workshop. He even knew some of the counterfeiting recipes, which the quarry workers kept secret and apparently improved from generation to generation.

'To increase the saleability of those faint imprints of petrifactions, whose colour can scarcely be distinguished from the colour of the stone containing them,' Karg wrote, 'the stone-cutters and lime-burners paint them with an infusion obtained by boiling the outer green shells of walnuts. They also know how to join broken pieces with deceptive verisimilitude by means of a whitish glue that they make out of powdered unslaked lime and sweet white cheese.'

The result was that the curators of the Karlsruhe Museum and of other scientific institutions had to subject the fossils from Öhningen to the closest examination, to determine what was genuine and what false. In their eagerness to purge their collections of retouched fossils, artificially prepared leaf-prints, and composite monstrosities, the outraged museum-keepers and collectors went too far. They threw the greater part of the forgeries into the refuse heaps. Hence only a few forgeries have been preserved to the present day as evidence of the skill of the fossil manufacturers of Öhningen.

In the meantime, however, another forger went to work. He was Leonhard Barth, a goldsmith and silversmith from the Swiss city of Stein am Rhein. Barth bought the quarries and set up a lively trade in fossils; contracts were made with Swiss and Dutch scientists, who were guaranteed continuous deliveries of fossils for very substantial remuneration. By about 1850, however, the yield from the Öhningen quarries was beginning to diminish. The fossil dealer bought another quarry belonging to the Grand Duke of Baden, and thus acquired a monopoly of all Öhningen fossils. In practice, that meant a monopoly of the stonecutters' recipes for forgeries. Soon Barth was supplying his customers with clever reproductions of fossils. The Teyler Museum in Haarlem, where Scheuchzer's famous 'witness of the Flood' rested, also received some remarkably artful counterfeits from Barth. Among them was a striking alligator-tortoise Barth composed from rudimentary fragments of six animals; the pieces had been put together with masterly skill to make a complete specimen. But this very masterpiece was to be his undoing. For as it happened, the custodian of the Teyler Museum, T. C. Winkler, was an expert on fossil tortoises. Winkler at once saw what was up. He sent the artificial tortoise back to Barth, took a thorough inventory of his collection to clean out all the counterfeits, and once more called the scientific world's attention to the forgers of Öhningen – not without paying a curious compliment to Leonhard Barth and his predecessors:

'The keenness, the inventiveness, and the artful ingenuity of the men of Öhningen knows no bounds.'

Charnel Houses Beneath the Earth

Until well into the nineteenth century the only explanation which the Christian churches would countenance for past geological changes and for the petrifaction of prehistoric organisms was the Deluge. Although the great age of geology and palaeontology was now opening, with these sciences at last outgrowing puberty, as it were, and advancing at a headlong pace, most savants were careful not to challenge the Biblical narrative as a source of information. The concept of the universal Deluge had given great impetus to the activity of collectors; it had also stimulated the writing of some first-rate books. By now, however, it was proving more and more of a drag upon research. Yet before the Deluge hypothesis gave way to other theories and approaches, a theologian was destined to discover the first genuine primordial man.

Among the collectors and the cave explorers of the time were a great many clergymen. This was only logical, for men of religion were particularly fired by the prospect of determining what witnesses of the Flood had looked like, and of uncovering physical evidence to support the Book of Genesis. Even afterwards, when the ideas of the Diluvians had long been refuted, the clergy continued dabbling in palaeontology. They turned their attention from details of the Old Testament Flood to the origin of man. Clerics such as the Abbé Bourgeois, Père Teilhard de Chardin, Hugo Obermaier, Abbé Henry Breuil, and many others made significant contributions to palaeontology and prehistory.

In the Cave of Gailenreuth

In 1774 one of the first of these clerical palaeontologists set about collecting fossil ivory in the cave of Gailenreuth, in that part of Bavaria known as 'the Franconian Switzerland'. To what extent Pastor Johann Friedrich Esper was an orthodox Diluvian is not known. At any rate, when he came upon all sorts of bones inside the cave, he did not immediately declare them the remains of drowned antediluvian animals. Instead, he indulged in some speculations that sound highly rationalistic: Had predatory animals buried their prey here? Had human beings used the cave as a carrion pit? These were virtually heretical ideas for a minister. Esper also differed from his contemporaries in the dramatic but extremely precise way he described his excursions into caves. He had a feeling for the peculiar atmosphere of grottoes, and vividly summoned up the emotions that overcome solitary explorers of such underground places:

Here you have not yet come to the end of the cavern. But the air is growing very stale. If you stay here a few hours, you can no longer make fire, even with the best match. Nor do I know what strange odour begins to permeate the clothing. . . . When you extract one of the bones and smash it, it often seems as if a repulsive and stunningly acrid smell enters the nostrils. . . . Nevertheless, no one who has even a little curiosity turns back without reaching the end.

Esper would make his way through the narrowest passage in caves, crawling on his stomach. Everywhere, he scratched bones and splinters of bones out of the clay. His lively fancy conjured up the ways in which the creatures whose remains he was finding must once have come to their deaths:

Did these hosts of creatures live in the vicinity? Did a single accident cause them all to die on this spot? Or what wonderful chance brought all their bones to these common burial places? I imagined their last moments; I thought of the dismal roars with which they breathed their last lamentations. . . . With some measure of horror I thought of the raging rebellion of such mighty bones against the accident that robbed them of life. . . . Now a great silence is imposed on all of them.

The more Esper probed into the cave, the more insistently did he pose questions. He spoke of accidents, of natural disasters; he raised still other conjectures and hypotheses on the origin of these great charnel houses. At first he thought that all the bones were human. It occurred to him that the cave of Gailenreuth might have been an ancient robbers' den where brigands had murdered and buried their victims. But after he had studied the bones more closely he realized that 'human beings with heads and teeth of such shapes were never seen on earth'. Those bones had belonged to predators – wolves and hyenas, he thought for a while. But had wolves and hyenas ever used the cave as a common burial ground? The idea seemed absurd.

After a while Esper realized that most of the animal remains in Gailenreuth cave bore a striking resemblance to the bones of bears. Esper called them 'Teutonic bears', conceiving them as brown bears such as exist to this day, whereas in reality the animals had been Ice Age cave bears. This error in no way detracts from Esper's achievement. Until his time the remains of cave bears were almost universally regarded as bones of dragons or other monsters. With the insight of genius, the Franconian pastor, though an arrant layman in anatomy and zoology, had made an analysis of the 'dragon bones' that came within a hair of the truth.

'Possibly a Deluge Man, Possibly a Druid or Christian'

A man who dared to ask such questions, and who entertained the possibility that events other than the Biblical Deluge might have led to the creation of fossils, would not fall into the usual cliché way of thinking when he suddenly encountered genuine, unquestionable human remains. Pastor Esper first discovered a human lower jaw in the clay of Gailenreuth cave, and then a human shoulder blade. After a long search he finally, 'with fearsome delight', dug up a well-preserved human skull.

An adherent of the Scheuchzer School would have instantly identified this find as a Deluge man. But Esper gave the matter some thought. The human remains lay side by side with unidentifiable animal bones. 'There they lie,' he wrote. 'Possibly the peasant and his beast, possibly the noble and his charger, possibly an antediluvian man, possibly a Druid or Christian. . . . How long have ye slept, ye children of the earth cradled in putrefaction? It seems as if the Creator Himself had laid you to rest.'

It is greatly to Esper's credit that he did reflect instead of piously accepting the current views of his time. In the end, it is true, he came back to the Flood. He wrote a book entitled *Detailed Account of Newly Discovered Zooliths*. It contained a great many plates with excellent reproductions of the Gailenreuth finds. Among Esper's comments on the bones in the cave were these striking sentences:

'There seem to have been great numbers of them. Never did the legions which conquered a continent leave so many remains of their existence. But who are they? It seems as if those floods which bore the Ark rolled these gravestones over them.'

'It seems . . .' Esper was still asking a question, suggesting a mystery. Perhaps he held with the ancient wisdom that doubt is the father of knowledge.

We do not know what human remains Esper excavated, whether Late Ice Age Cro-Magnon men or even Neanderthal men. For the human bones from Gailenreuth have been lost. But Esper had already given a valuable hint to anthropologists and prehistorians when he wrote: 'If human bones are also buried alongside the remains of antediluvian animals in the mud of this cave, man must have lived together with these animals in that period.'

By now Scheuchzer's 'Deluge man' was no longer taken seriously. The wind was beginning to shift. In France serious scholars were discussing the origin of various types of stone and were comparing the earth with other heavenly bodies. Early evolutionary hypotheses were being voiced. Bishop Ussher's dictum that the earth had been created some four thousand years ago had long since been outmoded; in Paris Count Buffon tried to demonstrate by experiment that the age of the earth must be at least seventy-five thousand years. In such a climate of tempestuous progress, no one was especially interested any longer in antediluvian man.

For that reason, Pastor Esper's book remained unnoticed. In 1804, it is true, Johann Christian Rosenmüller, the distinguished anatomist, once more examined the Gailenreuth cave and reminded the scientific world of Esper's discoveries. But to most scholars the clerical speleologist remained an ignorant dilettante. In his amateurish digging, it was assumed, he must have let human bones from an old cemetery slide into his bear's cave.

That was the end of the matter as far as official science was concerned.

In place of the Deluge theory there now arose a struggle between two scientific factions, each of which inscribed the name of a classical god on its banners. The matter at issue was the age-old question of which element is the prime cause of structural changes in the earth: water or fire.

Book Two

Neptune and Vulcan

Let the matter stand as it will, this I must say: I curse this beastly
lumber room of the new cosmogony; and certainly some bright young
man will come along who will be brave enough to oppose this crazy
consensus.

JOHANN WOLFGANG VON GOETHE,

circa 1830

A Theory of the Earth

Paris was the birthplace of modern natural science, and thus also the cradle of exact palaeontology. It had sheltered, then struck down the unfortunate Palissy, that genius of a potter who hit on the idea that fossils might be the remains of extinct organisms. At the time of Louis XV it had been the assembly point of the Encyclopedists, those rebellious writers, scholars, and philosophers who sought to gather all the knowledge of their century in a multi-volumed work and thus establish a 'sacred federation against fanaticism and tyranny'. Paris became identified with rationalism, which was beginning to prevail against the spirit of fanciful speculation and medieval superstition in the realm of knowledge.

But Paris, too, up to the outbreak of the French Revolution, was the site of a fateful institution which caused the scientific avant-garde much trouble: royal censorship. Palissy had fallen victim to the censor's verdict. The censor also proved the major foe of the French scientist who in the eighteenth century broke with all traditional views and arrived at pioneering insights into the past of the earth itself and the life that dwells on its crust. This creative spirit in palaeontology was Jean-Étienne Guettard.

As early as 1746 Guettard had put forward a number of bold ideas that ran directly counter to prevailing views. He prepared a geological map of France, compared geological conditions in northern France with those of southern England, and pointed out striking similarities between fossils in the two regions. 'This map,' he wrote, 'will show that a certain regularity prevails in distribution of varieties of rock, of metals, and of most fossils. Particular rocks and metals are not found indiscriminately in every region.'

Guettard concluded from this fact that no single Flood, but a number of different irruptions of the sea must have operated to shape the present aspect of the earth. In the following years he went into this question more thoroughly in a series of scientific publications. He studied the 'precipitous, rent, and heaped-up rocks' in eroded valleys and asked: 'What is the cause of this upheaval? Did it take place in ancient times? Is the cause operative from day to day? Is there only a single cause?'

The answer he found to these questions was revolutionary. The forces that had produced such tremendous changes were the same as those 'which we see before our eyes today': the waves of the sea, rainfall, wind, the flow of brooks and rivers. Naturally the Biblical four thousand years would not suffice to accomplish such transformations. But if you assumed that the earth were vastly older than theologians maintained, normal natural forces could account for even the most radical upheavals.

In advancing this thesis, Guettard was a direct forerunner of the actualistic geologists who in the middle of the nineteenth century abandoned the belief in global catastrophes and adopted the explanation of gradual change to account for the formation of the earth's features – thus founding modern geology.

Recantation at the Sorbonne

Guettard's theory of petrifactions was equally bold and novel. The oldest fossils, he had observed, lie in the deepest layers, the most recent in the topmost layers of the earth. This observation led him to the hypothesis that there have been a number of periods in the earth's history, periods differing in climatological and other respects, and that each period had possessed its own animal life, which later became extinct. He described the ancient nummulites, belemnites, and ammonites, the fossil whales and other creatures of the past. He dismissed the belief that the huge fossil bones were those of giants, and even had an inkling of the characteristic animals of the Mesozoic era, the saurians:

Naturalists are now able to identify those bones dug from the earth, which had been held to be the bones of giants. They know that these are skeletal parts of great fishes, of whales or of large amphibian quadrupeds which like hippopotamuses lived partly on land, partly in the sea or in rivers.

In those most ancient times there could not yet have been human beings on earth, Guettard maintained. 'Never yet have human bones been found which have undergone change in the earth and taken a more or less long step toward the petrified condition.'

All these theories and observations so negated the Biblical account of the Creation that the censor finally laid his hand on Guettard's writings. Guettard was summoned to the Sorbonne, where he had to defend himself before a committee of angry professors and theologians. Unless he publicly recanted his views, he would be forbidden to engage in further scientific work, he was told.

Jean-Étienne Guettard was no martyr. He bowed to the wishes of the Sorbonne and published a recantation. In extremely non-committal language he declared that he had not expressed himself quite correctly in regard to fossil fauna; that although no living counterparts to belemnites, ammonites, and archaic fishes and reptiles had been found as yet, this did not mean that they were necessarily altogether extinct. If looked for carefully, they would surely turn up sometime and somewhere in an obscure corner of the seas or the globe.

The Grand Seigneur and the Censorship

Other intellects of the first magnitude lived in Paris at the time and were adept at slipping through the meshes of the censor's net. One of the cleverest at this art was the most prominent French scientist of the age, Count Buffon. As superintendent of the royal gardens, Georges Louis Leclerc, Comte de Buffon, was virtually unassailable. He played the part of the elegant courtier and grand seigneur, and in this guise published the most audacious theories of cosmogony and evolution. His social position, together with his brilliant style, kept the censors baffled.

The censorship was always worried about Buffon. Some one hundred and twenty censors sifted through his *Histoire naturelle* and other scientific works. But the count's heretical views were so skilfully concealed behind his scintillating presentation that only seven of the censors discovered a few passages which might give offence. These were not enough to warrant a ban on the book. Worsted, the censors ceased to occupy themselves with the scholar-courtier. Thus in 1750 Buffon could venture to write a *Theory of the Earth* which went far beyond the hypotheses of Jean-Étienne Guettard.

The basis of this theory was Buffon's famous experiment with the two spheres of metal. He heated the spheres red-hot and from the time they took to cool, calculated the age of the earth at 74,832 years. Life could have existed on this planet for forty thousand years, Buffon estimated. Such a thesis broke with all traditional ideas and opened the way for the exploration of nebulously remote eras in the earth's past.

On fossils, Buffon wrote:

It may well be that a number of extinct species are among them. Shellfish, ammonites, and the strange fossil bones that have been found in Siberia, Canada, Ireland, and many other regions, seem to confirm this conjecture. For in our times we know of no animal to which such bones could be ascribed. Some are of extraordinary length and thickness. . . . Everything seems to suggest that they represent vanished forms, animals that once existed and today no longer exist.

From here it was only a step to the statement: All fossils are the remains of extinct forms. According to Buffon, there had not been one Creation but many, not one age of the earth, but many different periods in the earth's past, not one flora and fauna, but many such. Did this not imply that there had not been one Deluge, but many great catastrophes in the past history of the earth? And might not each of these catastrophes have wiped out the creatures of the previous epoch?

A few decades later this idea of catastrophes was to become the dominant theory in the schools. Buffon, however, did not himself accept it uncategorically; he regarded the history of the earth with more modern eyes than his successors:

Causes of rare, violent, and sudden effect need not concern us, for they do not belong to the normal course of nature. On the other hand, effects that take place daily, operations that go on repeatedly in an unending succession – these are the motive forces on which we must base our explanations.

This was exactly the same view Guettard had held.

In subsequent publications Buffon went a step further. His study of fossils led him to think that the earth must be far older than the calculations from his metal-spheres experiment indicated. In order to arrive at the age of a particular stratum of earth, he concluded, one had to determine the following spans of time: first, the time required for the development of its special fauna; second, the time in which these creatures had become distributed over their territory; third, the time in which they died out and turned to stone.

'We will then see,' he wrote, 'that the long stretch of seventy-five thousand years is not yet sufficient. The structure of all the great works of nature shows us that they can only have arisen by a slow succession of regular and continual movements.' He calculated that forty thousand years would be required to create a mountain a thousand feet high. Accordingly, the age of the earth would have to be reckoned in the hundreds of thousands if not the millions of years.

Buffon ventured no estimates. But he was fully aware that he had thrown open a door, and that henceforth other scientists would be confronted with an entirely new world. In 1787, the year before his death, he wrote in a testamentary letter:

These labours on the age of the earth alone would demand more time than will be granted me still to live; and I can do no more than commend my work to posterity. It is a pity that I must leave these interesting matters, these precious monuments of the past; but my own age will not allow me to examine everything and to draw from the material the conclusions of which I have glimmerings. Others after me will make the necessary calculations.

Buffon was right. Only twenty years later his disciple and successor, Jean Baptiste Lamarck, spoke bravely of 'boundless ages, thousands, indeed millions of centuries'. The new era had dawned.

Water, Fire, and Basalt

The eighteenth century was an age of scientific systematization. Following the example of Linnaeus, who in 1736 began sketching out his *System of Nature*, scientists in general commenced taking a general inventory of the natural realm. Botanists, zoologists, geologists, mineralogists, and palaeontologists realized that they could have no solid footing for their studies until the objects of their research were arranged systematically.

But as soon as a science has drawn up such a system, it asks itself questions: Why is nature arranged this way and not another way? And how had this arrangement come to be? Such questions had to be asked so that a rigid scheme could be converted back into a living science. After nature had been put in order, the task was to investigate how she had developed.

In Paris various clever and original theories were radically transforming previous conceptions of nature. In the German mining city of Freiberg, meanwhile, and in the Scottish capital of Edinburgh, practical research work was laying the basis for still deeper penetration into the secrets of earth's past. Their respective findings, however, differed materially. Two antagonistic scientific schools formed around the scientists in Germany and Scotland. Freiberg became a centre of worship of the sea-god Neptune, while Edinburgh paid homage to the god of fire, Vulcan.

The Ore Mountains of Saxony had long been a favourite haunt of geologists who did not concoct brilliant ideas and hypotheses at their desks, but instead followed the bidding of an Austrian scientist: 'Go out, buy sturdy shoes, climb the mountains, seek in the valleys, in the deserts, and on the coasts of the seas. Explore the remotest corners of the earth! Observe and experiment continually. For that is the way; there is no other for those who wish to understand nature.'

The Earth is a Child of Time

This was the principle followed by Abraham Gottlob Werner, Professor of Mineralogy and Metallurgy at the Freiberg Mining Academy. Werner came from a family of miners. At the early age of twenty-four he had written an important book, *On the External Characteristics of Fossils*. He next proposed to develop a system of rocks, minerals, and fossils. He was interested in the origin of soil, and realized that the crust of the earth does not consist of a haphazard accumulation of rocks, minerals, and types of soil, but that it shows as orderly an arrangement as the animal and the vegetable kingdoms.

Just as Linnaeus and his successors arranged plants and animals in phyla, classes, orders, and so on down to species, Werner arranged the earth's crusts into layers and formations. Every formation, he taught, consists of various strata lying one above the other. But formations are not always sharply delineated; there are transitional strata which often link them. 'The earth,' he concluded, 'is a child of time, and has been built up gradually.'

Many investigators long before Werner's time had observed or at least suspected as much. But Werner was the first geologist to distinguish, clearly and definitely, five major formations; and he was the first to realize that each of these formations corresponded to a different age in the earth's history. He was also the first to attempt a careful subdivision of the formations, taking into account the presumable age of the various strata. His system of stratification, which he published in 1787, became the basis for all later classificatory systems in geology.

Werner was less successful when he tried to answer questions of how and why. He believed that at some time in the distant past the earth's core, of which we can know nothing, became covered by a primeval ocean. This ocean contained in soluble form all the substances of which the earth's crust consists today. Gradually these substances precipitated out; they built up on the core, layer by layer, as scale builds up on the walls of a water container.

Thus, according to Werner, water was the cause of all geologic formations. His followers, who for a long period exerted great influence upon the development of geology, therefore called themselves Neptunists, after the ancient Roman god of water. In due course they became the objects of bitter attacks by the adherents of a scientific school who held just the opposite view, and who called themselves Plutonists, after the god of the underworld, or Vulcanists, after the god of fire.

The most prominent of the Plutonists lived and taught in Edinburgh; in addition he operated a farm in Berwickshire. His name was James Hutton, and he was a man of parts in the eighteenth-century manner – not only geologist and gentleman farmer, but also jurist, physician, physicist, and chemist. He invented a new sort of plough, established a chemical factory, explored the mountains and islands of Scotland, and even wrote on philosophy.

Hutton and his adherents, the Plutonists or Vulcanists, did not subscribe as thoroughly to the god of fire as the Neptunists to the god of water. They maintained that fire as well as water had contributed to the building of the earth's crust.

Hutton was deeply shocked when an article published by the Royal Irish Academy charged him with atheism. A pious Quaker, Hutton did not see that Werner's Neptunism might at a pinch be reconciled with the Biblical Deluge, while his Plutonism flatly contradicted Deluge theories. The accusation affected him so strongly that he fell severely ill. He rewrote his *Theory of the Earth*, thinking that he might not have expressed his ideas with sufficient clarity. But he never fully recovered his health, and died a few years later.

The Picture Book of Earth's History

At the beginning of the nineteenth century, every naturalist worth his salt belonged either to Werner's party or to Hutton's. Some advocated the Neptunists' primeval ocean, others the red-hot liquid interior of the earth postulated by the Vulcanists. While they feuded, a young surveyor was crawling around mine shafts and canals in England, collecting fossils and noting the differences among types of rock. This young man, William Smith, was to use this data to construct the first useful chronology of geologic strata.

Smith was completely self-educated. Born the son of a village smith in 1769, he on and off attended the dame school of his native hamlet. His father died when he was only eight years old. The uncle who took charge of him earned his livelihood making irrigation and drainage ditches for the farmers of Oxfordshire. William helped him with this work, learned how to draw, and at the age of eighteen attached himself to a surveyor who in his turn could scarcely read and write, but had a knack for building beautifully crafted compasses and measuring instruments. Smith learned all he could from the man, and henceforth called himself an 'engineer'.

Soon afterward Smith himself became a surveyor – a profession that enabled him to travel about the countryside a good deal. The different types of stone and stratifications of earth fascinated him. In every formation, he realized, there were different fossils. In fact a formation could be recognized by its *guide* or *index fossils*. This discovery of the guide fossil was to prove an extraordinarily fruitful inspiration.

At first William Smith knew hardly anything about fossils. Two clergymen of his acquaintance explained that they were the remains of marine animals. Nevertheless, he perceived at once, with the insight of genius, that fossils may be regarded as illustrations in the picture book of earth's history. They are the ideal clues to the character and the age of geological formations. As the German geologist Max Pfannenstiel has put it: 'By his fundamental discovery of guide fossils, Smith gave geologists the thread of Ariadne, the key to all geological research.'

The Father of English Geology

It is astonishing enough that a single individual succeeded in arranging in chronological order the formations and fossils of England. It is still more astonishing in the case of Smith because he had no training, no aid, no encouragement, and no financial support. Moreover, he had little gift for writing. Drawing was a different matter; he made the first geological maps of England and Wales. But he was not able to assemble his observations and discoveries into a book. Rather than sit at his desk, he much preferred tramping about the countryside, digging in the ground and making a nuisance of himself with quarry workmen and miners by badgering them about fossils.

The two clergymen who had enlightened him on the nature of fossils were themselves passionate collectors of petrified remains. They finally asked him to dictate his ideas to them. He agreed, and out of the collaboration came several books in which Smith analysed typical formations and described hundreds of fossils. To all intents and purposes he had already written the first thorough history of the earth and its fauna and flora.

The terms Smith devised for the various geological epochs and formations are for the most part no longer used today. Other names became established. The British geologists Sedgwick and Murchison introduced concepts such as Cambrian, Silurian, Devonian, and Permian; even during Smith's lifetime additional terms came into use: Carboniferous, Triassic, Jurassic, Cretaceous, Tertiary, Quaternary. For once Smith had shown the way, everyone fell to tabulating the ages of the earth and their characteristic sediments, rocks, and fossils. At every university the name of William Smith was known; his simple but inspired discovery had made him world-famous. He remained a simple man, however, and kept his distance from the specialists.

Membership in the London Geological Society was offered to him; he refused it on the grounds that 'theoretical geology is the business of a particular class of people; geological practice is the business of an entirely different class'. The soul of kindness, but deficient in manners, dressed in knickerbockers, looking not much better than a tramp, his jacket pockets always full of snuff – so a contemporary described the 'father of English geology'.

William Smith's discoveries proved to have considerable practical importance as well. His geologic maps gave valuable leads to mining magnates and industrialists. They showed where to look for the treasures of the earth, where best to dig wells, build canals, and establish mines. Had Smith offered to sell his knowledge to any of the coal or iron kings, he would have been able to accumulate a great fortune. But he never escaped from poverty. For as one of his friends put it: 'In his indescribable contempt for money he communicated his discoveries to anyone who sought information.'

Smith had at last opened the book of earth's history. It became apparent that the oldest fossils were also the most primitive, while the most recent represented more highly developed fauna and flora. Theories of evolution sprang up almost as a matter of course. Any thinker who reflected on the evolution of the earth found himself having to consider whether life, too, had evolved.

An Adventurer of the Mind

The earliest theories of evolution read like fairy tales, legends, or myths. Their contents are often marked by disarming naïveté. Their authors were not naturalists, but poets, philosophers, even statesmen and war heroes. Thus it is not surprising that specialists did not take these theories seriously.

In ancient Greece almost all the pre-Socratic natural philosophers believed that life had evolved. But just how they conceived of such evolution is difficult to say, for our information is fragmentary and usually has come to us at second or third hand, from informants who probably misunderstood the theories. Anaximander and other pre-Socratics apparently held that all life came out of water. Empedocles even offered a crude sketch of a theory of selection. But Aristotle delivered the deathblow to evolutionary ideas in natural philosophy. Instead, the dogma of the immutability of species became established. All living beings, it maintained, had been given their present form by a single act of creation. Transitions from species to species, let alone from lower to higher organisms, were impossible.

Like belief in the Biblical Deluge, the immutability of species became identified with Christian dogma. Anyone attempting to shake this faith might easily find himself in conflict with his church. For this reason, the few evolutionary theorists who raised their voices in the early stages of scientific development tried timorously to coordinate their ideas with the stories of the Bible.

How Big Was Noah's Ark?

The legend of Noah's Ark became a focal point of such efforts at coordination. According to the Bible, Noah on God's instructions had saved a pair of every species of animal from the Deluge. The more one might have doubts about the constancy of species, the more careful one had to be not to disparage Noah and his floating zoo. It occurred to some clever people that the legend of Noah might, however, be turned to use as proof that species must have evolved.

How big had Noah's Ark been? It must have been a ship of inconceivable size if Noah had managed to stow into it a pair of every species of animal. Was it possible to construct so huge a vessel? The first man to tackle this question was – as might have been expected – an experienced sailor: Sir Walter Raleigh, the great British admiral and explorer. Raleigh was imprisoned in the Tower shortly after the accession of King James, and remained there from 1603 to 1616. He whiled away the time writing a five-volume *History of the World* in which he set forth the idea that only the animals of the Old World could have been included in the Ark and thus saved by Noah. The animals of the New World, he suggested, must have evolved subsequently from these Old World species.

In 1685 Matthew Hale, Lord Chief Justice of England, turned his versatile mind to a consideration of the animals in Noah's Ark. Even the Old World species, he commented, were too numerous to have fitted into the Ark. Noah, he concluded, had been able to save only a few species – the original ancestors of present-day animals.

Hale's writings reveal the Lord Chief Justice as a forerunner of Lamarck and Darwin. He argued that all species were by no means created as we see them and that many species had changed greatly in the course of time. Robert Hooke, the great physicist, who also had an excellent knowledge of rocks and fossils went a step further. He maintained that in the course of the earth's history many species had been entirely wiped out, while others had varied and changed. He also believed that new varieties of some species had developed as the result of changing climates, soil conditions, and food. Thus Hooke must be credited with a first foreshadowing of the theory of environment.

Leibniz alluded to something of the same sort around 1700. He suggested that 'once upon a time, when the ocean covered everything, the animals which now inhabit the land must have been

89

marine animals; with the disappearance of this element they gradually became amphibia, and their descendants finally lost the habit of their original home'. But Leibniz was no scientific rebel; he did not want to lose the good opinion of theologians. Hence he quickly beat a retreat: 'But such arguments contradict the sacred writers, from whom it would be sinful to deviate.'

Fish Change to Birds

The most interesting and comprehensive of the eighteenth-century theories of descent was drafted by a French government official and travel writer, Benoît de Maillet. Maillet, a widely learned man, served as a diplomat in Egypt, Abyssinia, North Africa, and Italy. Around 1715 he wrote a book that attempted to break the chains of Biblical chronology by once more invoking the theories of the ancient Greek natural philosophers. According to de Maillet, the germs of the first living beings came out of space and assumed the form of marine creatures in a primeval earthly ocean. Some aquatic animals and plants accidentally came on land, could no longer flee back to the water, and therefore gradually adapted to life on dry land.

Benoît de Maillet described how fish could gradually develop into birds:

Then under the influence of the air the fins split, the supporting spines were transformed into quills, the drying scales into feathers. The skin was covered with down, the ventral fins became feet, the whole body assumed a different shape. Neck and beak lengthened, and at last the fish was transformed into a bird. On the whole, however, the kinship with the original shape remained; it will always be somewhat discernible.

Scientists and theologians who themselves believed in the most fantastic creation, deluge, and catastrophe theories derided the French diplomat who wished to change fish into birds by prestidigitation. But his critics failed to note that the author of this evolutionary theory was reckoning with very long spans of time. De Maillet did not assume that a fish became a bird in a few days or weeks. He, in the language of the Bible, regarded 'a thousand years as a day'.

In order to protect himself from persecution and his work

from difficulties, Maillet took certain precautions. He reversed the spelling of his name, producing the anagram Telliamed. Moreover, he attributes his theory of descent to an Indian philosopher, who presented it in an imaginary conversation with a French missionary. On top of all, he insured that his book would not be printed until eleven years after his death. During his lifetime, to be sure, many copies of the manuscript went the rounds of Parisian salons. Buffon read it and was much impressed; it is possible that 'Telliamed' stimulated him to develop his own ideas of evolution.

In the Beginning Was a Dot

Count Buffon's conception of evolution must be extracted from his writings with some pains; for the superintendent of the Royal Gardens always, as he put it in a letter, saw the censor before his eyes 'like an apparition'. In his *Theory of the Earth* there may be found the significant sentence: 'It can be assumed that all animals descend from a single living being which in the course of time, by perfection and degeneration, produced all other animal forms.' Buffon arrived at this idea because he had found everywhere in nature an 'original and general building plan'.

Buffon's great opponent, Linnaeus, also played with theories of evolution in the later years of his life. Linnaeus had prefaced his *System of Nature* with the categorical statement: 'There are as many species as existed at the beginning of the Creation.' But in a book entitled *Metamorphosis of Plants*, which he published fifteen years later, he had come around to a different view: 'The species of animals and plants, and the genera as well, are works of time; only the natural orders are works of the Creator.' At the end of his life Linnaeus professed the same theory as Buffon: 'At the beginning of existence was a small dot from which Creation proceeded and gradually expanded.'

Thus, in the one hundred and fifty years between Raleigh and Buffon, the idea of evolution had made amazing progress. What had initially been guesswork about the tonnage of Noah's Ark had by now become the starting point for a theory of evolution such as Lamarck and Darwin were soon to advance. Even philosophers regarded the idea as at least worth discussing.

Kant, for example, wrote in 1790 in his *Critique of Judgement*: 'The analogy of forms strengthens our surmise of a real kinship between them in their origin from a common ultimate mother.' Kant pointed out that palaeontologists were now faced with the great task of discovering in fossils the ancestors of present-day creatures, and in reading from the strata of the earth the course of life's development. He predicted that palaeontologists would 'initially excavate creatures of less serviceable form', but that in more recent strata they would find more and more perfect creatures, displaying ever greater variety, 'all sprung from one generative womb'.

Thus the time should have been ripe for a man of genius, one possessing a sound knowledge both of fossils and living organisms, with the courage to shape all these suppositions, hypotheses, and tentative ideas into a new, dynamic view of the world. Such a man did in fact appear on the scene. But oddly enough, although the ground was so well prepared for him, he failed to win acceptance for his views.

Possibly Kant had had a presentiment of the reason for this failure. He called the idea of evolution 'so monstrous that reason shrinks back from it, trembling'. It was all very well to toy with such ideas, but when they were seriously advanced they seemed far too perilous, 'daring adventures of reason', to use Kant's phrase. The age of evolutionary philosophies terminated in the story of the rise and fall of Jean Baptiste Lamarck.

New Conditions Create New Species

Ever since the Middle Ages the barons of Saint-Martin had been among the most powerful and respected families in the French province of Picardy. In the course of time, however, they had somewhat come down in the world, and Chevalier Monet de Lamarck, who belonged to an impoverished minor branch of the Saint-Martins, did not know how he would provide for his many children.

The sons of a Chevalier de Lamarck should have become professional army officers; so family tradition prescribed. But an officer's education was expensive, and there was not a sou to spare in the Chevalier's house in Bazentin. Consequently the father destined his eleventh and youngest child, Jean Baptiste, for the church. Jean Baptiste attended the Jesuit Seminary, but he did not have the temperament for an abbé. When his father died, he lost no time discarding his cassock, fought as a soldier in the Seven Years War, and rose to the rank of lieutenant. But soldiering was also not the right profession for Chevalier Jean Baptiste. At the age of twenty-four his health gave way and he had to put off the uniform and don civilian dress.

He went to Paris. For an impoverished young man with only a smattering of education, that was a bold step. Jean Baptiste Lamarck lived in the Latin Quarter, sporadically studying literature, music, medicine, and science, while earning his livelihood as a bank clerk. Then he became a tutor for a while, and at the age of thirty-four wrote a book on the flora of France. About this time he seems to have made the acquaintance of Rousseau, who probably encouraged him to continue his scientific studies.

Nevertheless, he would probably have remained unknown if Count Buffon had not been seeking a tutor and travel companion for his son. Lamarck applied for the job, was accepted (probably because of his *Flore Française*) and thus gained entry to the fascinating world of natural science in the home of the keeper of the Jardin du Roi.

Buffon quickly recognized that this poor fellow from Picardy was a good observer of nature and, above all, a talented writer. And since the naturalist count always needed competent assistants in the royal collections, gardens, and menageries which he administered, he gradually trained Lamarck as his assistant. The erstwhile seminarian, soldier, and clerk was first assigned to the botanical collection, then to the fossils in the collection of minerals, and finally to the lower animals in the zoological collection. This post also enabled Lamarck to marry at last. Before he was done, he was to marry four times and father seven children.

Jean Baptiste Lamarck's Struggle for Life

Buffon died a year before the outbreak of the French Revolution. His successor as keeper of the gardens and collections, the sentimental nature writer Bernardin de Saint-Pierre, whose novel *Paul et Virginie* had been something of a rage, was unequal to the scientific and administrative tasks of his job. During the turbulence of the revolutionary period the populace would probably have overrun the former king's gardens and plundered or destroyed the collection had it not been for Lamarck. While Saint-Pierre shivered in his boots, Lamarck bravely went to the Convention and persuaded the deputies to have the Republic take over and protect the precious heritage under the new name of *Jardin des Plantes*.

Soon afterward Lamarck became professor of zoology at the Musée Nationale d'Histoire Naturelle, as the Jardin was thereafter officially called. In the Bulletin of the Museum for 1794 he is identified in the following curious fashion: 'Lamarck – age fifty, married twice, wife pregnant; Professor of Zoology for Insects, Worms, and Microorganisms.' At an age when most men have long since reached the peak of their careers, Lamarck was barely at the beginning of his. He married for the third time, and it seemed as if his many marriages and children allowed him no time for any new and special accomplishments in science.

When the Academy of Sciences for reasons of economy cancelled a monthly subsidy he had been receiving, he addressed the following bitter plea to the government of the Republic: 'The loss of this pension and the enormous rise in the cost of foodstuffs have brought such misery down upon me and my numerous family that I have neither the necessary time nor the necessary mental freedom to carry out my scientific tasks in a useful way.'

Many such petitions have been preserved. Lamarck could not manage on his professor's salary, which was not surprising in view of the inflation. Several times he managed to obtain subsidies from a special fund of the Republic which was actually intended for poets and artists. But these grants barely allowed him to do more than keep his head above water. Even in the period of the Directory and of Napoleon, when he had a normal professorial income, he found himself in perpetual financial straits. He was a poor manager, and he unquestionably had too many children.

Nevertheless, during these years of personal distress and public turbulence, he succeeded in writing a voluminous natural history of invertebrate animals. That book made him one of the leading zoologists of France. His concern with the structure of invertebrates seemed to be arranged on a ladder, rising through progressively higher stages from polyp through worm to mollusc. On the last rungs of the ladder stood the still more highly perfected animals, the vertebrates.

Lamarck was all of fifty-seven years old when he issued his first repudiation of the principle of the immutability of species. He wrote a short study entitled *Recherches sur l'Organisation des Corps Vivants* in which he attempted – like Buffon before him – to trace all forms of animal life to a common origin.

'The most universal assumption that organisms form species which can always be distinguished by unalterable characteristics,' he declared,

belongs to a time when the sciences were as yet almost nonexistent. The more our knowledge has advanced, the more our embarrassment increases when we attempt to define a species. The more natural specimens are collected, the more obvious it becomes that almost all gaps between species are filled and our dividing lines fade away.

But if species are not unalterable, then the present-day animal world must have developed, as Buffon had already hinted, out of the forms which now lie in the earth in a fossil state. Lamarck had become an expert on fossils in the course of his apprenticeship at the Jardin des Plantes. In his lectures as professor of zoology, he had sketched a picture of a constantly changing earth whose living creatures are subject to incessant change from epoch to epoch:

Every cultivated person knows that nothing on the earth's surface remains forever in the same state. Everything undergoes in time the most various changes due to the shifting influences of sun, water, and other causes. . . . These shifting conditions alter the needs, the habits, and the manner of life of animals; and consequently the organisms themselves must change imperceptibly, even though these changes become perceptible only after considerable time has passed. . . . Among fossil remains are found a large number of animals whose analogous living forms are unknown to us. Why should they have become extinct, since after all man could not have exterminated them? Would it not be possible, on the contrary, that the fossilized individuals were the ancestors of still living and since altered species?

The Giraffe's Long Neck

In his *Philosophie Zoologique* Lamarck later presented the same idea even more clearly and definitely: 'Species pass into one another, from simple infusoria up to man. The fossil forms of organic life are the real, genuine forerunners of our present living beings.' Lamarck did not content himself with conjecture. He searched for a cogent explanation of the mechanism of such changes. New conditions, he thought, compelled animals (and plants as well) to adjust to new needs and habits. These new needs and habits in turn required new organs and abilities. Thus he arrived at the first of his laws of evolution:

In every animal that has not yet passed the peak of its evolution, an organ is gradually strengthened by frequent and continuous use; it is thus developed, enlarged, and acquires greater vigour proportionally to the duration of use. Consistent nonuse of an organ, on the contrary, imperceptibly weakens it, causes it to deteriorate, gradually reduces its abilities, and ultimately causes it to disappear.

Thus living organisms are the products of their conditions. This environmental theory could only hold, Lamarck realized, if the inheritance of acquired characteristics were possible. For there would have had to be countless generations, each perpetuating some minute alterations which represented a more adequate adjustment to the requirements of the milieu, before a new species would have developed out of an old one. On the basis of such reasoning Lamarck developed his second law of evolution:

'Whatever characteristic individuals may acquire or lose by consistent use or constant disuse of an organ is transmitted to the next generations – provided that the acquired changes are common to both parents.'

Lamarck attempted to explain these maxims to his students by vivid examples. He described how the short-necked ancestors of the giraffe kept stretching to reach the leaves of trees, so that the necks of succeeding generations grew longer and longer. Wading birds which go fishing on beaches or riverbanks needed to make their legs long in order not to get wet; in this way their legs became more and more stiltlike. Geese had to seek food in the mud at the bottom of ponds and so stretched their necks; in the course of many generations they acquired long supple necks. Moles, cave salamanders, and other animals of the darkness had no opportunity to use their eyes; consequently, their eyes shrank in size, or the animals actually went blind.

From all these facts Lamarck drew the conclusion that animals could deliberately be changed by exposing them to specific environmental influences or by constantly modifying their life conditions. If blind cave salamanders were exposed to the light for generations, their descendants would some day regain sight. If the tails of mice were cut off for generations, a race of tailless mice would ultimately arise among the descendants of the amputated animals. To the horror of his listeners, Lamarck proclaimed that it would also be possible to produce one-eyed human beings. All that was needed was to put out the left eye of children, marry these children to one another later on, and repeat the same cruel experiment upon their children and children's children. Ultimately human children would be born possessing only a right eye.

Lamarck did not, of course, seriously think of experimenting with human beings. He was an extremely humane man of tender sensibilities. Possibly he had chosen this example only because he suffered from an eye disease and lived in dread of going blind. And since he believed firmly in his theory, he feared that his children might have inherited his disease. For that reason he sternly insisted on eye exercises, for himself and his family, in order to strengthen and preserve their vision.

Lamarck did not hesitate to include man within his theory of descent:

If any higher species of ape were forced by conditions to abandon arboreal life, and if the individuals of this species throughout a long succession of generations were made to use their feet only for walking rather than for climbing, these quadrumanes would of a certainty gradually change to bimanes. . . . This new carriage leads to higher evolution of the senses, to development of the brain and of language. A species which perfects these abilities succeeds thereby in gaining dominion over other living beings, in taking possession of all likely territory and clambering to the highest stage of reason.

At that time no human or prehuman fossils were known, aside from Pastor Esper's neglected finds. In other respects, too, Lamarck's theory stood on shaky foundations, as we know today. His ideas on evolution were more philosophical than biological; he could substantiate neither the influence of environment on the form of living organisms nor the inheritance of acquired characteristics.

Another hundred years were to pass before the great quarrel between Lamarckians and Darwinists flared up over these questions. This dispute – between adherents of the environmental theory and adherents of the theory of selection – has continued raging to this day, and has long since become an intensely political controversy. For while the Western world had espoused Darwin's theory of selection, the East embraced Lamarck's thesis that man is a product of his conditions.

All the same, it was not these weak points in his doctrine which brought about Lamarck's fall. He came to grief over the very part of his theory which was the soundest – as science was later to discover: his belief that fossils represented the petrified ancestors of present-day organisms. The greatest zoologist and fossil specialist of those uneasy times declared war on him over this issue.

This man was twenty-five years Lamarck's junior. His name was Georges Cuvier.

The Lawgiver of Natural History

Historians of science have often been puzzled at the changes in fashion in scientific thought. Thus, in Paris at the time of Buffon it was almost a matter of course to believe in gradual changes in the earth's exterior. In Lamarck's time minds were open to the possibility of a similar gradual evolution of living organisms, yet a few decades later in the age of Cuvier the mode had changed in favour of annihilating global catastrophes, and evolution was no longer admissible. Perhaps Buffon, Lamarck, and Cuvier were typical representatives of their times. Not even the greatest geniuses among scientists and philosophers are immune from the influences of their periods. To be sure, it is individuals, not the ages, who turn the wheel of research, as Goethe long ago observed. But every individual is embedded in his society, in the prevailing conditions, and is willy-nilly formed and informed by the spirit of his age.

Buffon was a child of the Rococo. He was born in 1707 and died just before the outbreak of the Revolution. As a court scholar enjoying enormous prestige, he spent his life in the atmosphere of absolutism, had seen no violent changes in the history of his time, but only slow, scarcely perceptible alterations. Monarchs died and were replaced by their legitimate successors. The Encyclopaedists hoped to bring about intellectual progress by slow, persistent efforts of enlightenment. Politicians and diplomats exerted wit and cunning to secure this or that advantage for their countries or governments. And Nature seemed to utilize, for the advancement of her aims, much the same methods as men.

Lamarck, however, was a child of Classicism. He was born in 1744, was middle-aged by the outbreak of the Revolution, and had experienced both absolutism and the radical upheaval that followed it. Thus the idea had been forcibly impressed upon him that man could change fundamentally under the influence of new conditions. The Revolution had swept away all seemingly unalterable hierarchies, dogmas, and principles. Lamarck could scarcely help concluding that such constants did not exist in Nature either. The Empire and the Restoration which came later no longer made much impact on his mind; by the time Napoleon was toppled, Lamarck had become an old, blind, forgotten man.

Cuvier, finally, was a child of the Empire and the Restoration. He was born in 1769, was still a young man at the outbreak of the Revolution, and spent his life in extremely unstable times. The revolutionary regime gave way to the Napoleonic era with its wars; the period of the Empire was followed by the Bourbon Restoration. The people had risen, then fallen back again into their old condition; an emperor appeared like a comet and plunged to extinction like a meteor. Legitimate royalty was decapitated and raised its head again. Cuvier, at the end of his life a peer in France and President of the Council of State, concluded from the historical upheavals and cataclysms he had witnessed that the history of the earth was marked by catastrophes, and from his belief in the unalterability of human nature he drew conclusions about the immutability of species.

In contrast to Lamarck, Cuvier had been accorded the best possible education available for a young man thirsty for knowledge. His father, a pensioned officer of the Swiss Guard, lived in the Burgundian city of Montbéliard, which then bore the German name of Mömpelgard and belonged to Württemberg. The Cuvier family – the name had probably been Küfer originally – were strict Calvinists and felt more drawn to the Swabian-Swiss culture and way of life than to Gallicism. Montbéliard became a French possession and Cuvier a French citizen only after the victory of the Revolutionary armies.

Even in secondary school, Cuvier had been a passionate naturalist. He founded a scientific society in Montbéliard at the age of fourteen. The group of boys discussed a wide variety of philosophical and scientific subjects, and helped shape the town's keen interest in such matters. Cuvier's father wanted him to study Protestant theology in Tübingen. But his often pugnacious and insulting wit, which later on was to cause many of his scientific feuds, soon led to a row with the rector of the Tübingen seminary. He was expelled and had to look around for another field of study.

Luckily, Duke Karl of Württemberg had met the intelligent young man during a stay in Montbéliard. He gave him a stipend to attend the Stuttgart Military Academy, where Cuvier proved to be not so rebellious as his famous predecessor at the same institution, Friedrich Schiller. For four years Cuvier studied law, medicine, and natural history; he was regarded as one of the most gifted and promising of the students. Much as Schiller had done, he gathered a circle of students around himself. He, too, read to his fellows from his own writings – in his case not dramas but zoological notebooks.

Every Living Organism Forms a Whole

At the age of nineteen Cuvier had to give up his studies. Due to the Revolution his father's pension was stopped, and the young man had to look for work. He followed the example of many impoverished students and became a tutor. Since he had meantime discovered that his heart belonged to France, he took a position in Normandy. There, in Fiquainville on the shore of the Atlantic, he taught the children of a Count Héricy for six years. He seems to have enjoyed a great deal of freedom in the count's castle, for betweenwhiles he instructed the young physicians of the Fécamp military hospital in botany, assembled a large collection of marine animals and fossils, and studied all available works on natural science. Gradually he initiated a lively correspondence with the most important naturalists of the age.

While he was still in Normandy, the first glimmerings of his subsequently famous law of correlation came to him. 'The principle of correlation,' in the words of Emil Kuhn-Schnyder, professor of palaeontology in Zürich, 'holds that the various parts of animals do not exist side by side without relationship, but that they are closely interrelated. Every living organism forms a whole. No part can change without changes in the other parts. Consequently, the shape of the missing parts can be deduced from each separate part.'

This was one of Cuvier's most brilliant ideas. By now it is a commonplace of zoology, but at the time it was a startling insight. Animals with horns and hoofs, Cuvier observed, always have the teeth of herbivores. Animals with claws must have predators' teeth. Reptiles with solid rows of teeth are vegetarians, while on the other hand reptiles with fangs live on meat. 'First seek to obtain the teeth of the fossil animal,' Cuvier proclaimed, thus applying his law of correlation to fossils. 'In most cases you will be able to recognize simply by the molars whether the vanished creature was a herbivorous or carnivorous animal.'

Many fossils of vertebrate animals are found only in a shattered state, jumbled with other broken fragments. Until Cuvier's time, nothing could be made of such morsels. Collections were full of such incomplete fossils. Attempts to determine what species they represented were for the most part little more than guesswork. But if Cuvier were right about the law of correlation, it would no longer be so difficult to reconstruct the whole body of an animal from bits of bone or specimen teeth.

Later on, in Paris, Cuvier demonstrated by many practical examples that correlation of forms is indeed a meaningful principle of nature. By dint of years of work at the Jardin des Plantes and excavations in the basin of the Seine, he slowly arrived at a method of identifying fossils by means of his law of correlation. But the suggestions he had made in his letters to Parisian scholars while he himself was still only a tutor, sufficed to arouse the respectful attention of his correspondents.

One of them, three years younger than Cuvier, was already regarded by Parisian scientists as a youthful prodigy in zoology. At the age of twenty-one this young man had abandoned the study of theology and joined Lamarck and his fellow workers at the Jardin des Plantes. Since the Revolution was at that time devouring its children as well as its enemies, there were plenty of vacant teaching and administrative posts at the Jardin des Plantes. As a result, the intelligent former theology student became professor of zoology soon after reaching his twenty-first birthday. His name was Étienne Geoffroy de Saint-Hilaire.

Cuvier carried on an exceptionally lively correspondence with Saint-Hilaire. The ideas of both men seemed to supplement each other in ideal fashion. As early as 1794 Saint-Hilaire recognized Cuvier's talent and extended the hand of welcome: 'Come to Paris and assume, among us, the part of a second Linnaeus, a lawgiver in natural history!' These prophetic words were fulfilled: Cuvier in fact became the lawgiver of natural history and even more. As Napoleon set himself up as emperor and dictator of France and half of Europe, Cuvier became emperor and dictator of biology.

Having accepted the invitation and moved to Paris, Cuvier spent a year as a teacher of natural history at the Ecole Centrale of the Panthéon. Then Geoffroy de Saint-Hilaire and Lamarck managed to install him in an assistant's post at the Jardin des Plantes. Within another year Cuvier was accepted as the leading zoologist and palaeontologist of Paris. His name was by then known throughout Europe; at every university naturalists spoke of him with veneration.

Cuvier was the first comparative anatomist in the history of zoology. He distinguished the classes, orders, families, and genera of the animal kingdom by their internal organization rather than by external and often superficial characteristics. The same method, he found, could be applied to palaeontology. His law of correlation became the magic key; in fact, he developed it into the basic method of research and identification in palaeontology.

Worlds from Bleached Bones

The German palaeontologist Max Pfannenstiel has given an account of Cuvier's procedure for reconstructing fossil finds:

He had many teeth of herbivorous animals which had been found embedded in the gypsum of Montmartre. After he had assembled two rows of teeth, he perceived that they must have belonged to two species. One species had protruding canine teeth; in the second species the canines did not protrude beyond the crowns of the other teeth. ... Now Cuvier proceeded to reconstruct the heads of both species. Fortunately, he found fragments of skull with a few teeth still in their bony casing. Thus the relationship between the structure of the skull and the loose teeth was established.

Fragments of foot bone also corroborated Cuvier's sense that he was dealing with two different species. He found both three-toed and two-toed fore and hind feet. He then drew the logical conclusion that three toes would go with three fingers, two with two. It was a conclusion by analogy, based on principles of mathematical harmony. But which skull fitted which feet? The skull with the protruding canine teeth had some resemblance to a tapir's skull; the one with the smaller canines rather reminded Cuvier of a stag's head, although there were no antlers. Since tapirs have three and deer two toes, Cuvier associated the three-toed feet with Skull 1 and the two-toed feet with Skull 2.

One of the animals from the Montmartre gypsum was a *Palaeotherium*, a tapir-like perissodactyl; the second was an *Anoplotherium*, a hornless, deerlike artiodactyl of the same period, the Tertiary. Shortly after Cuvier had made his reconstruction, complete skeletons of both species were discovered. They confir-

The *Palaeotherium magnum*, a tapir-like odd-toed quadruped of the early Tertiary. (*After a sketch by Cuvier, 1810.*)

med the new 'lawgiver's' projection in all details. Cuvier himself explained what seemed his phenomenal prescience in terms of pure reason: 'No scientist who is able to think in analogies can fail to accept this conclusion. He must at once exclaim: This foot was created for this skull and that skull for that foot.'

Of another fossil, also from the Tertiary gypsum of Montmartre, he had at first only the teeth; the rest of the bones were still in the gypsum. Cuvier took advantage of this find to put on a dramatic demonstration for the students. He had never before dealt with such fossil teeth; the species was of a kind hitherto unknown. In the auditorium of the Collège de France he announced to his audience that from the teeth alone he could describe the animal as an opossum, a marsupial of the Early Tertiary. He would now chip the bones from the gypsum and guaranteed that he would find the typical pouch bone of the marsupials.

Marsupials exist today only in Australia and South America, except for the opossum of North America. To everyone in the auditorium Cuvier's prediction seemed totally wild. Montmartre could never have been inhabited by opossums or other pouched mammals. But Cuvier set about chipping the gypsum away from the bone with every sign of total self-confidence. The students watched tensely, as if witnessing a piece of detective work. At last the skeleton lay exposed, with lo and behold the predicted pouch bone. Cuvier's further explanations were drowned out by a thunderous ovation such as had seldom filled the amphitheatre of the venerable Collège de France.

By feats like these, Cuvier soon won the reputation for infallibility. Honoré de Balzac was a young man at the time of his great excavations and explications of fossils. A passage in Balzac's *The Wild Ass's Skin* testifies, more forcefully than any specialist's tribute, to the impact of Cuvier upon the thought of his time, and suggests the reason he inevitably won out over his rival Lamarck:

Is not Cuvier the great poet of our era? Byron has given admirable expression to certain moral conflicts, but our immortal naturalist has reconstructed past worlds from a few bleached bones; has rebuilt cities, like Cadmus, with monsters' teeth; has animated forests with all the secrets of zoology gleaned from a piece of coal; has discovered a giant population from the footprints of a mammoth. These forms stand erect, grow large, and fill regions commensurate with their giant size. He treats figures like a poet: when he sets a zero beside a seven, it produces awe. He can call up nothingness before you without the phrases of a charlatan. He searches a lump of gypsum, finds an impression in it, says to you, 'Behold!' All at once marble takes an animal shape, the dead come to life, the history of the world is laid open before you.

On one occasion, when Cuvier was already the most famous scientist in Paris, high-spirited students ventured to play a joke on him. One of them dressed up as the devil. Equipped with the traditional horns on his forehead and hoofs on his feet, the young man broke into Cuvier's bedroom at night, crying: 'I am the devil and have come to eat you.'

A group of students crowded against the windows anticipating a bit of drama. But Cuvier calmly regarded the masked monster and smilingly shook his head. 'Since you have horns and hoofs,' he replied, 'you are completely harmless. By the law of correlation, you must feed only on plants.'

The *bon mot* quickly circulated among the students of the scientific faculty, who thenceforth twitted the theology students with sardonic remarks about the devil, who had been exposed, according to Cuvier's system, as a completely harmless herbivore.

Cuvier Against Lamarck

At the time Cuvier came to Paris, young Geoffroy de Saint-Hilaire as well as Lamarck were extremely hospitable to the idea of evolution. The twenty-one year old Geoffroy had as good as taken his stand for evolution in his inaugural lectures: 'I emphatically doubt the constancy of species.' Geoffroy, like Lamarck, had made studies of fossils, and the conviction had gradually formed in his mind that the geologic ages constituted the steps on a ladder leading from the lowest to the highest creatures.

Geoffroy de Saint-Hilaire practised comparative anatomy in much the manner of Cuvier. But he drew altogether different conclusions from his data. He compared the fins of fish with legs, air bladders with lungs, reptilian forelegs with avian wings. He searched for similarities, whereas Cuvier looked for differences. Both procedures are indispensable in the biological sciences – the former in order to establish the unequivocal earmarks of species and orders, the other in order to uncover structural and evolutionary linkages.

Cuvier, however, refused to grant the equal validity of these two approaches. He regarded the ideas of evolution as a philosophical notion, and he hated speculation. All his researches, excavations, and anatomical studies had revealed to him a neatly ordered world in which every species, genus, order, and class was distinguished from every other by clearly recognizable characteristics. He had found no transitions from form to form. Every geological formation also had its unambiguous characteristics; Cuvier saw no evidence for the assertion that one geologic age slowly and gradually passed into the next.

Of course Cuvier observed, as had Lamarck and Geoffroy de Saint-Hilaire, that animal life in the earliest ages of the earth must have consisted of extremely archaic and primitive forms. From epoch to epoch the creatures seemed to grow more modern and to approach more closely the forms of present-day fauna. But Cuvier did not conclude from this that life had gradually evolved from lower to higher forms. Instead he took refuge in a hypothesis which seemed to fit very well into the revolutionary spirit of his own age: the theory of geological revolutions.

'The animal kingdom,' he maintained,

was organized in earlier ages on the same principles as today. Among the various theories on the origin of organisms there is none more improbable than the one that the families of animals have arisen as a result of step-by-step evolution. ... If this hypothesis were correct, traces of such step-by-step transformations would have to be found; but up to the present there has not yet appeared a single indication of that. Why have the bowels of the earth preserved no evidences of any such genealogy? Because it did not exist, because the species in earlier times were just as constant as they are today.

Catastrophes Demand Victims

According to Cuvier, the earth was shaken many times in its history by upheavals and cataclysms. Some of these were local events; in such cases they only annihilated life in the affected regions, so that new living beings drifted in from other regions and repopulated the devastated areas. That was Cuvier's explanation for the astonishing fact that fossil elephants, rhinoceroses, tapirs, and opossums could be found in the soil of France. Once upon a time the animals had been wiped out by catastrophes; immigrants from other unscathed regions subsequently replaced them.

Some of the cataclysms, however, had proved much more drastic. In his principal work on palaeontology, *Recherches sur les ossements fossiles*, Cuvier stated:

Life on earth has been frequently interrupted by frightful events. Innumerable organisms have become the victims of such catastrophes. Invading waters have swallowed up the inhabitants of dry land; the sudden rise of the sea bottom has deposited aquatic animals on land. Their species have vanished forever; they have left behind only sparse remains, which the naturalist is currently striving to interpret.

Cuvier did not exactly state that life on earth had been totally extinguished and re-created by God in superior form; that theory remained for his successors to promulgate. Cuvier himself was more ambiguous: 'I by no means assert the necessity of a new Creation for living species; I say only that they did not live in the places where we now find them. They must therefore have come in from other regions.'

Cuvier did not go into the question of how new species had arisen from such invasions. He was a rationalist, but at the same time he was a Christian, a convinced Protestant. The idea that God could have created new forms after every sizable geological upheaval, or even a whole new fauna, may therefore have been not too far from his mind. He may well have taken it for granted that God had made each new creation better than the preceding one, which was why organisms appeared increasingly modern and perfected from age to age.

But he said nothing about these matters. His theory of geological revolutions and animal immigration stood on the shakiest foundations. Biographers of Cuvier are apt to emphasize that he himself was not to blame for the ideas of his successors – the doctrine of successive cataclysms and new creations – which dominated the thinking of naturalists until the time of Darwin. But this dodges the issue. The only alternative to the catastrophe theory was some sort of evolutionary theory. If Cuvier had thought differently from his disciples, he would necessarily have defended a theory of evolution instead of battling it with all his might.

Even so great a zoologist as Cuvier did not yet dare to excise the story of the Biblical Deluge from the history of our planet. He regarded the Flood as merely the last of the great geological revolutions. 'It carried the largest land animals to the northern polar regions, where we now find their cadavers embedded in ice, with skin, hair, and flesh well preserved.'

Cuvier did not believe in the existence of man in primordial times. But since the Bible represented man as well established on earth before the Flood, Cuvier cautiously added to his dictum 'L'homme fossile n'existe pas!' a few explanatory comments: 'To be sure, this is not to say that from four to six thousand years ago, in other words before the last catastrophic Deluge, men could not yet have existed. Perhaps their dwellings plunged into abysses and their bones sank to the bottom of the present seas. . . . There remained only that small number of individuals who reproduced our race.'

Every Age has its Fauna

Cuvier's anatomical and palaeontological feats were so dramatic that his contemporaries failed to notice the obvious weaknesses and vaguenesses in his catastrophe theory. Just as his law of correlation and his system of comparative anatomy became established, his view of periodic geological revolutions likewise acquired the force of law. All the fossils being found at the time seemed to confirm his views. For the first of the giant saurians were now being turned up, as well as such weird-looking primitive creatures as *Dinotherium*, a swamp-dwelling proboscidian, and *Megatherium*, a sloth as big as an elephant. It seemed impossible that such monsters could be the ancestors or primordial relatives of present-day species. Thus the great progress being made in the discovery and examination of fossils tended to drive the theories of Lamarck and Geoffroy de Saint-Hilaire more and more into the background.

Cuvier adopted the (now outmoded) geological terminology of Giovanni Arduino, which divided time into four successive eras, and assigned to each its characteristic fauna:

In the oldest Era, which he called *Primary Time*, sea lilies, giant crabs, ammonites, and belemnites had flourished. They died out as the result of catastrophic geological upheavals.

The next era, *Secondary Time*, was the age of fishes and saurians.

The characteristic animals of the next age, the *Tertiary*, according to Cuvier, were the tapir-like *Palaeotherium*, the stag-like *Anoplotherium*, and huge animals such as *Dinotherium* and *Megatherium*; a tremendous irruption of the ocean spelled their end.

In *Quaternary Time*, finally, the most recent era, present-day fauna appeared, including the apes and man. Some of these, such as the mammoth, mastodon, woolly rhinoceros, and giant deer, fell victim in the course of the Quaternary to the diluvium, the last great geological catastrophe – in other words, the Biblical

Flood. But others lived through the great deluge and engendered the animals of the present geological age.

As we have said, Cuvier did not quite hold that his catastrophes had radically extinguished all life each time. Land animals, he thought, would have suffered more from the geological upheavals than aquatic creatures. While Neptune or Pluto could wipe out whole classes of land animals with water or fire, the inhabitants of the seas would be in a position 'to preserve themselves in a few quiet places, from which they would be able to disperse anew after the ocean had settled down again'. In fact, Cuvier did indeed consider that marine animals might undergo some degree of evolution:

When we study the various marine deposits, we recognize that the composition of marine landscapes was subject to constant changes. It is understandable that such changes could not leave the animal inhabitants unchanged. Species and genera changed along with the particular formations. By such variations the aquatic animals have been led step by step to their present state.

This agreed, basically, with the theory expressed in Lamarck's *Philosophie Zoologique*. But Cuvier insisted that it was applicable only to marine creatures – and then only by way of exception. He certainly did not conceive of a general evolution of all life. His attitude towards the two theoreticians of evolution at the Jardin des Plantes was highly subjective and ambivalent. In spite of many spirited disputes, he was a close friend of Geoffroy de Saint-Hilaire. His relationship with Lamarck, on the other hand, ranged between cool and icy. As long as Saint-Hilaire was associated with his older colleague, Cuvier refrained from too harsh criticism of the idea of evolution and from any public attack upon its advocates. But Geoffroy did not remain in Paris. He took part in Napoleon's Egyptian campaign as an officer in the French army and scientific commissioner. In Egypt he collected countless fossils in the Nile valley. When the campaign came to its inglorious end, Geoffroy insured that the palaeontological and archaeological collections were shipped home to France. Napoleon rewarded his bravery by appointing him scientific *inspecteur* of the Iberian Peninsula.

Unfortunately, Geoffroy had no time, in Spain, to develop his theories further. In Paris, however, Cuvier and Lamarck now confronted one another eye to eye – two men who had long since become bitter scientific enemies.

Fall of a Rebel

By now Lamarck had grown almost blind; he would be led into the lecture room by his favourite daughter, Cornélie. His lectures, moreover, were poorly attended; his ideas about evolution were not as spellbinding as Cuvier's demonstrations. 'Citizens,' Lamarck began one of his lectures, 'you must always proceed from the simplest to the most complicated; you thereby have the Ariadne's thread which runs through the entire organic world! You will thereby obtain a precise picture of progress in nature. And you will thereby also be convinced that the very simplest organisms gave rise to all other living beings.'

Conspicuously failing, he did not help his cause by repeatedly citing the example of the blind cave salamanders. Another favourite theme of his was the pitiable caged birds which after five or six years of confinement lost their power to fly. If such birds were constantly crossed with one another, and their descendants likewise locked in tiny cages, he contended, after many generations a new species of birds, incapable of flight, would arise. The whole problem of the loss of function in important organs preoccupied him more than he may have realized.

At the same hour, Cuvier was speaking in an adjoining, jam-packed lecture hall about 'the crazy new fashion', as he called Lamarck's theory. If one accepted Lamarck's arguments, he quipped, then 'time and circumstances alone will suffice to change an infusorian or a polyp into a frog, a swan, or an elephant'. Bitingly, he added: 'A system of nature which is based on such assumptions may possibly stir the imagination of a poet; but it cannot for a second withstand examination by a specialist who has ever studied a hand, an internal organ, or merely a bird's feather.'

More and more students deserted Lamarck for Cuvier. Once Cuvier actually entered Lamarck's lecture room and challenged him outright. By asserting that fossil varieties are the predecessors of contemporary organisms, he cried out with deliberate bellicosity, Lamarck had attacked his lifework – and he must regard this as a personal affront. He therefore called upon Lamarck in the presence of the assembled students to prove his thesis or to formally withdraw it.

Lamarck, of course, could not furnish the required proof. The sudden assault threw him into total confusion. The students, of course, found their professor's humiliation a huge joke. Then Cuvier himself mounted the platform, delivered a brief but polished lecture on his theory, supporting every detail with precise drawings of the geological periods, rock formations, fossils, and reconstructions he mentioned. Then he delivered the final blow to poor Lamarck.

'All those who can see,' he cried,

must perforce accept my theory. Everyone recognizes the clear, immutable structural plans which govern the organization of different groups of animals. One organ collaborates meaningfully with another; all the functions of animals stand in reciprocal relationships to one another. How could something entirely new and different evolve out of such meaningful collaboration? Such a wild claim could only be made by someone who is blind – blind to the facts of Nature.

The insult struck home. Lamarck visibly winced. He attempted a reply:

According to that point of view, a lion would eat meat because he has tearing teeth, claws, and a predator's stomach. I see it otherwise. It is not that certain organs cause a certain mode of life. The reverse is the case: the mode of life, the habits, and the conditions of the environment are what have created the forms, organs, and peculiarities of animals.

'If that were true,' Cuvier replied brutally, 'then one would have to conclude that you, Monsieur Lamarck, have made too little use of your eyesight. By your own theory, you yourself are to blame for your blindness. Had you looked at nature more, instead of philosophizing, you might have prevented your disability.'

This debate, which has been transcribed in a number of different versions, seems to have sounded the knell for poor Jean Baptiste Lamarck. Before long he had virtually no students left, and his contemporaries ceased to pay him any heed. Soon total blindness forced his retirement to private life. The economic distress which had dogged him all his life now became even more acute. He occasionally gave private lessons; but after his fourth wife died, he sank deeper and deeper into misery.

Nevertheless, he summoned up the strength to complete the two last books of his eleven-volume *Natural History of Invertebrate Animals*, dictating entirely from memory to his two daughters. Sometimes he would have Cornélie lead him to the Jardin des Plantes, so he could listen to lectures. Cornélie literally went begging in order to procure their food. He died in 1829, at the age of eighty-five. He was buried in a pauper's grave on Montparnasse – one of those anonymous graves turned over after five years to make room for the corpses of other poor people. No one knows what became of his mortal remains.

Geoffroy de Saint-Hilaire delivered a kindly memorial address in the name of his colleagues. Cuvier, however, insisted on composing an obituary of his own. It was so stinging, and added up to so vindictive an attack upon the dead man, that everyone was shocked. The Académie Française, which published a memorial volume, refused to include Cuvier's speech. This was the second scandal Cuvier had stirred up in connexion with Lamarck – and this time he did not emerge victorious.

Two years later, by which time Cuvier was a peer of France and on the verge of being appointed Minister of the Interior, he at last had his chance to deliver a speech on Lamarck before the Académie Française. His tone was somewhat more temperate, but still grudging and discourteous. The dictator of biology could not yet forgive his dead antagonist. On this score, the greatest of French naturalists makes a poor showing in the eyes of history.

The Monster

Cuvier's views seemed to be brilliantly corroborated when the first fossils of saurians were discovered. These Mesozoic lizards appeared to be so monstrous and strange that their existence could not possibly be reconciled with the idea of evolution. Out of what could they have developed? And what creatures could possibly have descended from them? Since only a very few saurians were known in Cuvier's time, it seemed as if they were the ideal evidence for his theory. The evolutionarily significant saurian groups, the ancestors of present-day reptiles, birds, and mammals, were not discovered until much later.

The first land saurian to enter the annals of science sidled in, as it were, through the back door. In 1706 a young man named Spener, son of the famous Pietist Philipp Jakob Spener, found a curious fossil near Suhl in Thuringia, which he thought must be a crocodile. The fossil was not particularly large. Twelve years later a similar specimen was found. When it was submitted to Cuvier, he straightway identified it as an extremely archaic reptile, superficially resembling a giant lizard. The Greek for 'lizard' is *sauros*; Cuvier therefore called the fossil *Proterosaurus speneri*, 'Spener's early lizard'. Thus the term *saurian* made its entry into palaeontology. This special proterosaur flourished, as we now know, in the Permian, the last period of the Palaeozoic. Hence it is one of the oldest saurians, in terms of descent.

The Stolen Saurian

The second saurian that came Cuvier's way was a rather alarming-looking monster. In contrast to *Proterosaurus*, it had had a distinctly dramatic fate. But it, too, had come to light by sheer chance.

In the early period of palaeontological research most fossils were unearthed not by systematic digging but were stumbled on in the course of mining or quarrying, deep ploughing of a field, or road building. In lucky cases the find would fall into the hands of an engineer or workman who was curious or intelligent enough to inform the nearest specialist. Frequently that was not the case, and in the course of time many valuable specimens undoubtedly vanished without a trace, or leaving at most a few rumours and vague legends behind.

There was one spot in Holland, Petersberg, near the city of Maastricht, which even in prehistoric times had been a favourite source of flint for weapon and tool making. Later, the Romans dug underground shafts in the area to obtain limestone. Even in the eighteenth century the labyrinths they had dug continued to yield fine quality white writing chalk. Fossils were frequently found in this chalk, especially fossil sea urchins, which were prized in the locality as amulets.

One of the quarries was owned by a clergyman named Godin. One day in 1776 a former German army surgeon, Dr Hoffmann, called on Canon Godin. Hoffman was collecting fossils for the famous Haarlem museum of the Dutch patron of science, Pieter van der Hulst Teyler, and wanted permission to dig for fossils in the chalk quarry. Godin, assuming that Hoffmann was thinking chiefly of the well-known fossil sea urchins, consented.

Hoffman had not been digging long when he came upon a gigantic animal skull of a totally unknown type. The monster had fearsome teeth, and to judge the size of the body by the size of the skull, must have been between thirty and forty feet long. Naturally Hoffman was delighted with this extraordinary find. He had the skull chiselled out of the chalk at his own expense, and took it to Haarlem. There Pieter Teyler at once consulted the most noted anatomist of Holland, Peter Camper, on the scientific identification of the Petersberg monster.

At first Camper thought the animal must be an archaic whale. The account he published of the find created a sensation. Even Canon Godin heard of the matter, and was furious that so valuable a piece had been taken from him by what he regarded as trickery. Although he knew nothing about fossils and had no collection of his own, he rushed to court and sued the Teyler Museum, demanding the return of the skull. Since he was formally in the right, he won the suit and brought the petrified monster back to Maastricht. There he stowed it in his house and gave short shrift to all the scientists who came hoping to study the specimen.

French palaeontologists questioned Camper's interpretation. One of them, Faujas de Saint-Fond, thought the monster must be a crocodile. But the matter could not be solved until someone was able to examine the skull again.

Cuvier had studied the accounts of the Petersberg case with particular care. At this very time French revolutionary troops under Pichegru were marching across the Dutch borders; in 1795 they stood outside Maastricht. Cuvier made arrangements with General Pichegru and with People's Deputy Freicine, who was to take over the civil administration as soon as Maastricht fell to the French troops. The general gave orders that Godin's house, with its precious scientific treasure, was to be left unharmed. Godin seemed hardly grateful, for he went to the trouble of having his skull hidden somewhere else in the city in the dark of night. Freicine found this out as soon as the city fell. He issued a proclamation to his soldiers promising six hundred bottles of wine to the 'second discoverer of the skull'. With such a reward in view, the soldiers ransacked the town, and next morning twelve grenadiers came to the Deputy triumphantly bearing the skull and claiming their payment.

Ever since, the Petersberg skull has remained in the Jardin des Plantes in Paris. Cuvier immediately recognized it as the skull of a gigantic lizard. In honour of Hoffmann he named it *Mosasaurus hoffmanni*. This second use of 'saurian' for an extinct reptile helped establish the term in popular as well as scientific use.

The mosasaurs were, as Cuvier recognized, sea dwellers and mighty predators. Later finds have shown that they lived over most of the earth in the seas of the Upper Cretaceous. Their kinship with other marine saurians is very remote; their closest relatives are the giant snakes of the present day.

Fighting Dragons

Cuvier had the opportunity to make the acquaintance of some members of the dinosaur order, although the name 'dinosaur' was not coined until long after his day. In 1824 William Buckland, the English geologist and theologian, described the bony remains of a reptile which subsequently turned out to be a predatory lizard that walked on its hind legs. With its clawed hands, which had developed into veritable grappling irons, the megalosaur must have been a fearsome creature. He could rip open the bellies of large, herbivorous saurians with these claws of his, following up with his mighty, dagger-toothed jaws, and was thus probably more than a match for the largest reptiles of the Cretaceous.

The animal that was presumably this predator's principal victim is also classed among the dinosaurs and came to light at the same time and at the same sites in England. A country doctor named Gideon Mantell, a native of the town of Lewes, south of London, was wont to spend his free time walking in Tilgate Forest and searching in the chalk deposits for fossils. On one such outing in 1822 Dr Mantell's wife found a few odd-looking teeth. Mantell could not identify them and sent them to Cuvier in Paris. But Cuvier seems not to have been in his best form at this time. After some delay he identified the finds as the incisors of a fossil rhinoceros. In the meantime the Mantells had discovered very peculiar petrified footprints which were evidently those of an enormous three-toed animal. At first they did not associate the footprints with the teeth. But presently Mantell dug up a number of bones that obviously came from an animal of the same species as the teeth. Cuvier thought they were hippopotamus bones.

Finally it occurred to both Mantell and Cuvier, independently, that the mysterious animal from the lower Cretaceous strata of Tilgate Forest must have been a large reptile. The petrified footprints, too, as it turned out, belonged to the same monster. Mantell compared the fossil teeth with those of present-day iguanas and imagined – wrongly – that there were certain resemblances. He therefore called the giant reptile *Iguanodon*.

The discovery of the iguanodon inspired a spate of highly imaginative saurian reconstructions. Mantell wrote his popular *The Wonders of Geology*, and wanted a picture of an iguanodon for the frontispiece. The artist John Martin obliged with such a portrait, which turned out to be that of a short-legged dragon with a massive tail. When the famous Crystal Palace Exposition opened, its garden was adorned with sculptures of gigantic antediluvian creatures. Waterhouse Hawkins, the sculptor, had also represented the iguanodon as a clumsy four-footed dragon. Fifty years were to pass before the world learned that the animal did not walk on all fours, but erect on two legs, and that it fed peacefully on the shoots of coniferous trees.

In the meanwhile other terrestrial and marine saurians had been dug up. A number of them came from different geological periods and thus could never have met in their lifetimes. But this was not yet known. The catastrophe theory of recurrent geologic upheavals fostered the idea that these primordial creatures had lived in an atmosphere of violence, preying on each other and fighting to the death. Thus the paintings and reconstructions of the time showed Jurassic saurians locked in combat with Cretaceous saurians. Iguanodon and hylaeosaur, ichthyosaur and plesiosaur, mosasaur and megalosaur grappled, open-mawed, with one another. The marine lizards, who in reality had had fins and could no more move about on land than whales and dolphins, were pictured clambering out on the shore and attacking dinosaurs; the dinosaurs in their turn lumbered into the water to strike at ichthyosaurs. The age of saurians was conceived in the most bloodthirsty terms.

The novelist Jules Verne was greatly taken with these representations of battles among the dragons of the ancient world. In one of his science fiction novels, he introduced a scene of just such a life-and-death struggle between two saurians. For who could have believed that the saurians were other than ferocious beasts?

In truth, only a few of them were. Most of the terrestrial saurians were innocent eaters of plants and small animals; the marine saurians lived on fish and molluscs. Among the marine saurians were two very different types which were destined to catch the fancy of the general public even more than *Mosasaurus* and *Iguanodon*. These were *Ichthyosaurus* and the *Plesiosaurus*. They, too, were discovered, described, and reconstructed during Cuvier's lifetime.

It can no longer be determined who came on the first ichthyosaur, or when. The black Jurassic shales of Swabia, Franconia, and England are so full of ichthyosaurian remains that it would be a miracle if some of their bones had not been turned up centuries ago in the course of building or quarrying.

In 1669, at any rate, a Welsh antiquarian named Edward Lloyd published a picture of such a monster from a one-time Liassic ocean in a richly illustrated book entitled *Lithophylacci Britannici Ichnographia*. Lloyd thought the ichthyosaur a large petrified fish, though a fish of a special sort. He reasoned that fish eggs had been drawn into the clouds when water evaporated from the sea, and then fallen on dry land during rains. In the sand and soil they developed not into living but into 'stone' fish.

Forty years later, as we have seen, Scheuchzer and Baier quarrelled over the nature of the ichthyosaurian backbones they had found in Altdorf. Another forty years later, in 1749, the first complete ichthyosaur was found near Bad Boll, Swabia – where so many have been found since. The finder's name was Mohr; he was a practising physician in a nearby town. The good country doctor must be forgiven for having identified the animal as a shark; saurians were then wholly unknown.

Mohr's 'shark' rested unnoticed in the natural history collection of a Stuttgart Gymnasium until 1824. Then it was seen by another physician, Georg Friedrich Jäger, who recognized at once that it could not be a fish. Jäger was another of those amateur geologists and fossil collectors who swarmed over southwestern Germany at the time. He betook himself at once to Bad Boll, where people told him that such creatures frequently came to light in the slate quarries.

Upon subjecting Mohr's find to closer examination, Jäger saw resemblances to skeletons, skulls, and bones which had been discovered along the coast of southern England. These bones had been discussed for many years by the greatest authorities of England and France; they constituted the well-known 'Lyme Regis mystery'. Jäger wrote a little book in which he offered a drawing of the skeleton of Bad Boll and commented that there was no need for anyone to travel to England to probe such 'mysteries'; the mountains of Swabia were full of them.

Mary and the Fish-Saurians

There is a romantic story behind the finds of Lyme Regis. A man named Richard Anning had opened a souvenir shop at the bathing resort of Lyme Regis. The British were just beginning to discover the healthful effects of sea-bathing, and hitherto insignificant towns along the seacoast were overnight turning into populous spas. Among the merchandise Anning sold pretty sea shells and other flotsam – the kind of thing offered to this day in booths and shops along seashores. His twelve-year-old daughter, Mary, helped gather the shells.

Every so often collectors of natural objects would turn up among the visitors. Anning noticed that petrified shells were more in demand and fetched higher prices than the ordinary kind, even though the latter were embellished with the motto: 'In Memory of Lyme Regis'. Consequently, he sent little Mary out to dig fossil mollusc shells out of the Jurassic strata along the coast.

One day – this was in the year 1811 – Mary came upon the

skeleton of an ichthyosaur. Of course she had no idea what kind of creature she had found, but she recognized at once that it must be a rarity. As luck would have it, just at this time a prominent personage interested in natural history happened to be staying in Lyme Regis. He was Sir Everard Home, physician to the king and professor of anatomy and surgery at London University.

Sir Everard bought the ichthyosaur. And while Mary Anning continued to dig for fossils, with all the more zeal because of the high price, Great Britain's leading surgeon pondered over his acquisition. Was it an amphibian, a reptile, or a whale? In 1819 Sir Everard at last decided to dub the 'Mystery of Lyme Regis' an amphibian. He felt it might be related to the blind cave salamanders found in the karst of Slovenia; at this time, this animal had just been described scientifically. The salamander bore the scientific name of *Proteus*; Sir Everard therefore baptized the creature of Lyme Regis *Proteosaurus*.

Meanwhile, however, Mary had dug up more of these animals. The British Museum bought some of them. William Daniel Conybeare, a geologist who had accumulated a large collection of fossil stones in Cardiff, also obtained a skeleton. He came to Lyme Regis to study the site at which it had been found. Mary, by this time twenty-two years of age, had grown into a professional dealer in fossils and expert on saurians.

George Koenig, the mineralogist of the British Museum, encountered Conybeare there. It occurred to both of them that the creature of Lyme Regis had been half fish, half reptile. And since, to judge by Mary Anning's numerous finds, it had been fairly common in its lifetime, Koenig and Conybeare named it *Ichthyosaurus communis*, 'the common fish-lizard'.

Soon afterward Dr Jäger informed the world public of the Stuttgart ichthyosaur and of the other sites in the Swabian Jurassic strata where it had been found. In Swabia, too, a number of persons began digging for fossils as a profession, and even opened shops dealing in 'bones in stones'. The Swabian Jurassic, it turned out, was far richer than the Jurassic of southern England.

Engaged in the fossil trade at the time were pastors, peasants, village physicians, quarry owners, and even a nobleman, one Count Mandelsloh zu Urach. Bitter feuds often sprang up among these dealers in 'dragons'. They were given to paying secret visits to their rivals' sites; they attempted to buy the best pieces away from each other. Finally they were driven to a more or less equitable division of territories. Some maintained agencies in various towns and villages; others, veritable 'travelling salesmen in fossils', periodically visited universities, museums, and the great foreign collections. A descendant of the Mohr who had discovered the first Swabian ichthyosaur in 1749 hit on the brilliant idea of opening a fossil business in the United States. He earned a fortune exporting specimens to America.

Lyme Regis, meanwhile, continued to be a rich site. The indefatigable Mary Anning discovered the first *Plesiosaurus*, that extraordinarily long-necked marine saurian which subsequently was to be connected with the legendary sea serpent and mysterious monster of Loch Ness. (Plesiosaurs also turned up by and by in the Jurassic formations of Swabia and Franconia, but not so commonly as the ichthyosaurs.) In 1828 the capable Englishwoman made still another coup when she dug up an excellently preserved flying saurian.

Flying Dragons

Towards the end of his life Cuvier had a sufficient number of saurians at his disposal to form some notion of the appearance and habits of these ancient reptiles. He quite correctly identified the plesiosaurs as large inhabitants of shallow bays. Swimming on the surface of the water, they lived chiefly on molluscs and fish. With their small heads, serpentine necks, and large steering fins, they were something of an oddity in nature. Fossil remnants must have been known early, for they occur in medieval drawings, where their fins are represented as wings and they themselves as 'flying dragons'.

On the other hand the real flying dragons, the *Pterosauria*, appear in ancient pictures as swimming and diving aquatic creatures. The first discoverer of a pterosaur, the Italian writer and naturalist Cosmo Alessandro Collini, tagged them as such. Collini, onetime secretary to Voltaire, in 1784 described a flying dragon about the size of a pigeon which had been found in the lithographic slate of Solnhofen in Franconia – later a world-famous palaeontological site. Collini described the creature as 'an unknown amphibious marine animal of dubious zoological classification'.

Soon afterwards Collini took charge of the Mannheim Natural History Collection, and installed the pterosaur there. The slab of slate from Solnhofen soon became the most controversial fossil of the age. The professors could not agree. Johann Friedrich Blumenbach of Göttingen said that the bones were those of a water bird; Johannes Wagler of Munich, specialist in anatomy and reptiles, maintained that it must have swum in the ocean like present-day marine tortoises. Other scientists said that it was a prehistoric bat or an 'epicene creature, half bird and half bat'.

Cuvier alone hit on the right interpretation. Sent to Germany in 1813 as a special envoy of Napoleon, he found time in spite of his political business to visit the natural history collections at Mannheim. With unerring sharpness he realized that the pigeon-sized animal in the slab of slate could only be a flying reptile. He named it *Pterodactylus*, 'flying finger', because the animal's fourth finger had become enormously long in order to extend the flying membrane.

But these conclusions by no means put an end to further and fantastic reconstructions. The French scientist de la Beche depicted the pterodactyl as literally a winged dragon straight out of a book of fairy tales. Hawkins and Martin, attempting a pictorial rendering of Mary Anning's pterosaur, tried to hedge: from their drawings the peculiar creatures might equally well have lived in the water or in the air. These efforts at illustration looked, as Othenio Abel has remarked, 'more like the visionary phantoms of a Breughel in hell than a scientific reconstruction'.

As fantastic as these drawings were the contention by some naturalists and natural philosophers with an evolutionary bias that saurians had not really died out, but had merely changed into certain groups of animals still living. Birds, they argued, had developed out of pterosaurs, dolphins out of ichthyosaurs, baleen whales out of the mosasaurs, and sperm whales out of the plesiosaurs. Such wild speculations ended by discrediting the idea of evolution. Rather than visualize such transformations, both laymen and scientists would sooner believe that the ancient dragons of land, sea, and air had been extinguished in some universal catastrophe.

Cuvier always held aloof from such fanciful notions. He tried to interpret every fossil as objectively and correctly as possible, eschewing any part of the stories of primordial dragons and monsters so loved in the nineteenth century. His supremacy was such that he continued to dominate palaeontology decades after his death. Thus, it was Cuvier's fault that palaeontology kept marking time until the second half of the nineteenth century. Excellent scientists followed in his footsteps: the brothers d'Orbigny in France, Louis Agassiz in Switzerland and the United States, Richard Owen in England, Hermann von Meyer in Germany. But for all their feats in their own specialities, they did not go beyond Cuvier in general theory. On the contrary, not only Cuvier's accomplishments but also his errors ruled the theory, practice, and teaching of palaeontology in a more and more exaggerated fashion. The science of palaeontology suffered the same fate as the giant saurians: it drifted into a blind alley in its evolution.

'A Significant Incident'

Few naturalists have ever risen to such high positions in politics as Cuvier. Even before Napoleon he had been Inspector General of Public Instruction – that is, a kind of Secretary of Education in France. Under the Empire he was assigned the task of re-organizing universities and academies on the French pattern in the occupied territories. Then he served as Counsellor of State and Imperial Special Envoy for the territories on the left bank of the Rhine. As such, his task was the somewhat delicate one of preparing the populace for a general uprising in case the Allies should invade.

The Restoration proved no hindrance to his career. Louis XVIII appointed him Chancellor of the Sorbonne very soon after Napoleon's fall. In 1818 he was slated for the post of Minister of the Interior, but refused it because he did not want to interrupt his scientific studies. Instead, he devoted his spare time to the cause of French Protestants. He persuaded the government to set up fifty new pastorates, and was appointed overseer of the Protestant theological faculty of Paris University. Then he was made a baron, became President of the Council of State, Member of the Académie Française and finally, under Charles X, peer of France. He had just received his second appointment as Minister of the Interior, in the government of Louis Philippe, when he died suddenly on 13 May 1832.

A man who has managed to survive many political systems, winning the highest honours in them all, comes in time to feel himself infallible in his science. Even during Cuvier's lifetime, teeth and remnants of the fossilized bones of human beings were turning up in various parts of the world. At the time the mammoth had been dug up at Cannstatt, in 1700, a fossil human skull had been found. In Thuringia a respected German palaeontologist, Baron Ernst Friedrich von Schlotheim, came upon human teeth and thighbones in association with the bones of extinct animals. At Lahr in Baden the great geologist Ami Boué dug human remains out of the loess. In a cave on the island of Cerigo one of the most celebrated biologists of that age, Lazzaro Spallanzani, found fossil human bones. Whereupon Johann Friedrich Blumenbach, the 'German Cuvier', declared: 'There is no conceivable reason why fossil human bones should not be found in the top levels of our globe, just like the fossil bones of elephants and rhinoceroses.' But Cuvier rigidly insisted on his thesis that fossil human beings did not exist.

The first French cave explorations were made in the first quarter of the nineteenth century. From 1821 to 1826 Paul Tournal, d'Hombres Firmas, and de Christol discovered human bones in several caves of southern France. Similar finds were made in England. A cave on the Welsh coast yielded a completely preserved human skeleton along with stone tools and other traces of human activity. On the strength of this, William Buckland, the leading geologist of Oxford, eagerly declared himself a Diluvian. He sharply attacked Cuvier, remarking that the so-called 'great authorities' could no longer honestly maintain that geology offered no proof of the account in Genesis.

Cuvier stuck to his guns: '*L'homme fossile n'existe pas.*' Disdainfully he replied to Buckland that of course he believed in a great flood which had taken place during the last geologic revolution; but he did not believe that human victims of that flood would ever be discovered. Those who had drowned had sunk to the bottom of the ocean; the few individuals who had escaped 'spread and increased on the new surface of the earth, and their descendants founded settlements and erected monuments; they are now collecting the facts of natural history and thinking up scientific systems'.

There really was no good reason for Cuvier to have taken this line; for it did not affect his theory in the least whether remains of the unfortunate victims of the Flood rested forever in the depths of the seas or could be found in convenient caves and quarries. His objection was more psychological than theological; he feared that if he conceded a single drowned witness of the Flood, the protagonists of the 'new kind of craziness', the evolutionists, would at once make an animal man or an ape man out of it.

Combat in Paris

But though Cuvier did his best to quell it, the principle of evolution rose once more to plague him. Goethe described the affair in the 'Yearbooks for Scientific Criticism' of 1830:

At a meeting of the French Academy on February 22 of this year an important incident occurred which cannot fail to have highly significant consequences. In this sanctuary of science a dispute has arisen over a scientific point which threatens to become personal, but which when examined closely goes far beyond personalities. What herein comes to light is the persistent conflict between two modes of thinking into which the scientific world has long been divided but which is now coming to a crisis.

The two modes of thinking, according to Goethe, were personified by Cuvier and Geoffroy de Saint-Hilaire: 'Cuvier labours indefatigably to make distinctions, to describe precisely the material at hand, and he has been winning mastery over an immeasurably broad realm. Geoffroy de Saint-Hilaire, on the other hand, seeks quietly to discover the analogies of creatures and their mysterious relationships.'

But what exactly had happened? Geoffroy had presented a 'theory of analogies' based on the example of molluscs and fishes, which held that all animals are subject to a common, unitary structural design. Cuvier rapped back with characteristic sharpness – and with an incidental jibe directed at Goethe, whose fondness for natural philosophy he knew well: 'I am well aware that in certain minds there may well lie concealed behind this theory of analogies, at least quite confusedly, another and very old theory which has long since been refuted [Cuvier was thinking of Lamarck], but which is again being fetched out by certain Germans in order to promote that pantheistic system which they call natural philosophy.'

The disputes between Cuvier and Geoffroy de Saint-Hilaire were spun out throughout 1830. Geoffroy's *unité de composition organique*, which Goethe found so congenial, really meant: all organisms pass into one another; there is in nature none of those abrupt gaps that Cuvier stresses. The disputants focused their arguments upon, in Goethe's words, 'those curious fossils to which scientists are only nowadays directing their attention'. Geoffroy was able to catch Cuvier out in mistakes in his own special field.

The problem fossil was that of a primordial 'dragon' which Cuvier refused to recognize as an ancestor of present-day reptiles – the *Teleosaurus*. Bones and parts of the skull of this saurian had been found in the Upper Lias of southern England. Geoffroy now asserted, in opposition to Cuvier, that the teleosaur had been a primitive crocodile. Cuvier, he added, had described the teleosaur skull carelessly, and in part incorrectly, so that the coincidences with crocodile skulls had not been perceived.

Can we imagine nowadays that the public would turn out in droves to attend a scientific lecture in which the occiput of a saurian was being compared with that of a crocodile? Yet this is what happened in Paris at that time. Such was the crush at the crocodile disputations that Cuvier – who abhorred crowds – bowed out of the public argumentation in October 1830 and contented himself with continuing the discussion on paper. Incidentally, Geoffroy had been right: the teleosaur is a Jurassic ancestor of the crocodile.

Mass Slaughter and New Creation

Goethe's hope that Geoffroy's approach would prevail in the future was not fulfilled. After Cuvier's death his disciples consolidated the hold of the catastrophe theory and the dogma of the immutability of species upon the scientific world. In geology Elie de Beaumont, a descendant of the great Anglo-Norman family of the Beaumonts, described the history of the earth as a colossal drama of volcanic eruptions, rendings, shrinkings, and other violent events. In palaeontology Alcide d'Orbigny put forward his doctrine of 'successive creations', taken straight from Cuvier. As late as 1849 d'Orbigny could write:

A first creation appears at the Silurian stage. After its total annihilation by some geological cause and after the passage of a considerable span of time a second creation took place at the Devonian stage. Thereafter twenty-seven successive creations repopulated the entire earth anew with plants and animals, geological upheavals having each time destroyed all living nature. These are certain but incomprehensible facts. We shall content ourselves with noting them without attempting to penetrate into the metaphysical mystery that surrounds them.

Even after the appearance of Darwin, d'Orbigny and his numerous followers continued to refer to Cuvier and would not retreat an inch from their far-fetched doctrine of catastrophes and new creations. Meanwhile the palaeontologists had long since determined that the characteristic animals of particular geological times frequently occurred in other epochs. On that score alone, all life could not have been extinguished at the end of one geological period and created anew at the beginning of the next. But d'Orbigny would not be moved from his orthodox view: 'If an animal form appears in two different geological periods, even if it is identical in the two periods, we must nevertheless assume that it became extinct in between and was afterwards created anew. Therefore it is a new species, even though not distinguishable from the earlier one.'

During the decades between Cuvier's death and Darwin's victory, the catastrophe theorists held almost all the key academic posts. Whatever new data were gathered about primordial life they reinterpreted in their own terms. Every collector of fossils saw fossilized corals, molluscs, and other marine animals 'lying in their original position', as one catastrophe theorist put it, 'all cheerfully flourishing and reproducing, but not dying out and atrophying'. Was that not a decisive argument against the intervention of catastrophes? On the contrary. The scholar in question – the noted German geologist Friedrich Hoffman – regarded this fact rather as proof of his theory: 'The terrible catastrophe must therefore have been the work almost of a moment. Animals and plants did not gradually die out and give way to other forms; they perished suddenly; the conditions they required for existence suddenly ceased to obtain.'

While the idea of evolution continued to be regarded as some sort of crank notion, or a whimsy of romantic natural philosophy, sober scientists and laymen everywhere in the civilized world preferred the far more improbable hypothesis that the flora and fauna of our globe have changed radically between twenty-five and thirty times in the course of the ages. God or an 'unknown natural force,' it was argued, had periodically committed mass slaughter of all organisms and then, after total destruction, hastily recreated animals and plants.

Just at this time, in the middle of the nineteenth century, it turned out that Cuvier's 'last geologic revolution', the Diluvium, had in reality not been a sudden, totally destructive flood, but a succession of glaciations. But the very man who brought the idea of the Ice Ages to the fore, Louis Agassiz, most obstinately upheld the orthodox theory of creation.

'There is no identity at all between fossil and living species,' Agassiz maintained. 'From the geological point of view there is no direct connexion between two different geological epochs. Every epoch has its own fauna.'

Nevertheless Louis Agassiz inadvertently overthrew this rigid dogma in banishing the Biblical Deluge from the history of the earth. In its stead there was now the Ice Age. And with the discovery of the Ice Age a new era of geology and of palaeontological research began.

Book Three

The Discovery of the Ice Age

Death entered with his terrors; with one blow of his violent hand he destroyed a mighty creation; he wrapped all nature in a shroud.

LOUIS AGASSIZ, 1841

'An Epoch of Great Cold'

When partisans of the catastrophe theory were asked for tangible proof of their theory, they were apt to point to the erratic blocks and masses of scree which lie scattered over plains as if dropped or pushed there by giant hands: the moraines. *Errare* means 'to go astray'; and these worn fragments of rock, sometimes as smooth as if they had been polished, seemed literally to have strayed. Obviously they belonged to the mountains. But to the bafflement of geologists they were found with particular frequency in lowlands, far from mountains. How had they come there?

The first person to set about systematically solving this enigma was a Swiss: Horace Benedict de Saussure. He came from a family of wealthy Genevan patricians, and is honoured by the city of Geneva as one of her greatest sons; for he was not only the first geologist to investigate the Alps, but also the founder of alpinism. Saussure became a professor of philosophy at the age of only twenty-two. Later, he took a statesman's part in the history of his country as a member of the Genevan Council of Two Hundred, but kept up his scientific work as well. His *Essai sur l'Hygrométrie* (1804) is considered one of the classical works of meteorological research. But he was not satisfied with only intellectual pursuits. He tramped through all the alpine regions. In 1787 he became the first man to climb Europe's highest peak, Mont Blanc. Five years later he conquered Monte Rosa. He explored all the passes, collecting characteristic stones and plants, and from 1779 to 1796 wrote an eight-volume study, *Voyage dans les Alpes*, which dispelled all prejudices against the hitherto shunned and dreaded high mountains, the *montagnes maudites*.

In the foothills of the alpine regions, Saussure repeatedly saw the picture familiar to everyone acquainted with the moraines of the lowlands: everywhere he found rounded, polished erratic blocks, masses of rubble, deposits of scree and gravel. To Saussure's great credit, he described these mysterious collections of boulders and stones accurately. As yet he had no idea where they had come from or what elementary forces might have transported and polished them – although, as a climber familiar with the alpine glaciers, he ought to have recognized what he could have seen on any expedition in the mountains.

Saussure concluded, however, that the erratic blocks were testimony to some geological catastrophe, a *débâcle*: 'There is no doubt at all that these stones were moved by water. . . . But where could such quantities of water have come from? What gave them such violent momentum? How could they have carried boulders up heights which are separated from the Alps by wide and deep valleys?'

The ocean, Saussure concluded, had once covered a large part of the mountains. Suddenly, a powerful earthquake must have torn open gigantic hollows. 'The waters rushed with frightful violence into these abysses; they carried with them enormous quantities of earth, sand, and fragmented stone. This great mass of water mingled with stone caused the depositing of the materials whose remnants we still see lying today.'

For the present no one offered any better explanation. Leopold von Buch, one of the most important German geologists in the age of Cuvier, followed in Saussure's footsteps in the Alps and studied the erratic blocks of the North German plain. North German drift boulders and terminal moraines, he suggested cautiously, might have come from Scandinavia, 'although their route and the way in which they arrived at their present position remains a mystery'. He too, like Saussure, cited 'one of the catastrophic floods of Switzerland' as the explanation for the moraines in the foothills of the Alps.

Other scientists spoke of tremendous periodic irruptions of the ocean which had cleaved the high mountain range into innumerable separate mountains and ranges. 'All the boulders, the enormous fragments from the Alps, were rolled down by the power of the sea and spread over the splintered valleys.' Thus the good old idea of the Deluge was cropping up in a new scientific guise; the geologists simply could not shake it off. It took the offer of a scientific prize to break the insidious habit.

The Factor of Time

In 1848 the Royal Society of the Sciences in Göttingen, at the suggestion of the zoologist Blumenbach, offered a prize for the scientist who could provide 'the most thorough and comprehensive investigation of the changes in the earth's surface'. The Göttingen group was hardly bent on overthrowing prevailing doctrines; in fact, the contest called for an exposition of ancient geological revolutions. But the prize went to a man who believed in neither geological revolutions nor catastrophic deluges. He was Karl Ernst Adolf von Hoff of Gotha, a high government official in the small Thuringian duchy of Saxe-Coburg-Gotha.

Von Hoff was the editor of a local geological journal. He had written several treatises on the geography of Thuringia, had exchanged scientific data with Goethe, and corresponded with Alexander von Humboldt on the nature of basalt, a subject much discussed in the early part of the nineteenth century. For a time he served as legation staff secretary and university inspector. Later he became head of the observatory, curator of the art collections, and presiding officer of the Church Council of Gotha. Here then was a man at home in diplomacy, in the Church, in art, and in science – one of those many-faceted personalities who flourished at the time in these small German duchies and especially in Thuringia. But the name of von Hoff might never have been known beyond the boundaries of his native land, had it not been for the prize offer from Göttingen. He was almost fifty years old when he embarked on the problem set by the prize offer.

The result was a work in three volumes. The long-winded title scarcely sounds inviting: *History of the Natural Changes in the Surface of the Earth Demonstrated by Tradition.* Neither von Hoff nor the Göttingen prize judges could guess that the book was to mark the beginning of a new era in the history of geology: the era of actualism, of evolution.

In the first volume, von Hoff described the eternal battle of the sea against dry land, with many historical examples. He also provided a keen analysis of the Atlantis legend, which was rather in vogue at the time (as it was to be repeatedly in the years to come). Von Hoff firmly relegated Atlantis to the realm of fable. In his second volume he dealt with the effects of earthquakes and volcanoes, in the third volume with the changes produced in the crust of the earth by wind, water, ice, and organisms.

From all his data he drew the conclusion that there could not have been violent catastrophes in the history of the earth. Hitherto, he argued, scientists had been reckoning with far too small time spans. Changes in the physical appearance of the earth naturally seemed like revolutionary upheavals when they were compressed into a few centuries or a few millennia; but they became slow and peaceful developments 'if they are given a long enough time' – that is, hundreds of thousands or millions of years.

'Neither tradition nor the observation of nature offer proofs for a single or repeated upheaval or for the destruction of an entire organic creation,' von Hoff summed up. 'Decisive reasons not only allow but require us to ascribe the changes which have been observed or are still being observed on the earth's surface to still operative forces. ... The immeasurable vastness of the spans of time in which these forces have worked gradually and continually suffices to explain the phenomena and changes of ancient time.'

To the holders of catastrophe theories, von Hoff retorted: 'In the history of the globe it is not at all necessary to economize on time, but definitely on energy.' He took strong issue with the current hypotheses of a universal Flood. Later he wrote a fiery polemic against Dean Buckland, the English head of the Diluvian party.

Von Hoff had a fresh theory about the erratic blocks and the mysterious moraines. They had not been deposited by violent masses of water. Intuitively he recognized what scientists familiar with the Alps, such as Saussure and von Buch, should have guessed from casual observation of glaciers: 'Perhaps these fragments were transported in and on ice.'

When he made this suggestion, von Hoff was not thinking of a glaciation of North and Central Europe; he did not have an 'Ice Age' in mind. Rather, he assumed that the boulders had been carried to their present sites on the drift ice of the sea and of alpine lakes. There had been intimations of this drift theory already: back in 1786 a mine administrator named Voigt in Weimar had suggested that ice could be 'a means for bringing stones and other loads from one bank of a river or lake to the other'.

Even the idea that not drift ice but glaciers might move huge masses of rock had already come up. In 1802 the mathematician John Playfair suggested that only a glacier could 'bear rocks on its surface and transport them over great distances'. That was not, of course, a glacier theory such as was developed by the later investigators of the Ice Ages. But it was an idea that should have started geologists thinking.

But no one had paid attention to Voigt or Playfair. The response to von Hoff's work, however, was much greater – although it remains questionable whether the Gotha scholar would alone have succeeded in revolutionizing geology. Luckily he did not have much of a fight on his hands. For now events began to move swiftly. In England, in Switzerland, and in Weimar, within a single short decade, the same ideas were presented. It was as though von Hoff had only had to press a button to release, all over Europe, powerful intellectual pressures which set the existing structure of dogma swaying.

But the boldest, most interesting, and most modern theory of the erratic blocks came not from a geologist but from a poet.

'For All That Ice We Need Cold'

About the same time that von Hoff was awarded the Göttingen prize, Goethe learned that large ice floes had brought blocks of granite across the Danish Sound. He at once thought of the notorious erratic blocks and noted, in keeping with von Hoff's notion of drift ice: 'To move them we cannot do without the ice.' In another jotting, the poet who was also privy councillor, collector of minerals, and erstwhile director of mining operations, sketched the method by which boulders might be carried by glaciers. He called this idea one of his 'favourite thoughts', and applied it to the alpine moraines in the vicinity of the Lake of Geneva:

'I let the glaciers drop on and on down through the valleys to the edge of the lake; upon them the detached blocks slide on a smooth sloping surface and are pushed along, as still happens today. They remain lying by the lake, the ice melts, and we find them there to this day.' In this way, Goethe thought, the erratic blocks of the North German plain could have ridden on the backs of glaciers from Scandinavia.

Goethe's antipathy for the violence of the catastrophe theories probably led him to return again and again, during the last years of his life, to this idea of the forces of ice. Glaciers pushing masses of rubble before them, polishing boulders and depositing them in valleys and plains, seemed to him more in accord with the natural order of things than calamitous floods, volcanic eruptions, earthquakes, and other cataclysms. 'Where you, gentlemen, only stir tumult and would give us accounts of frightful uproar,' he writes in a fictional disputation outlined around 1818, 'everything runs along silently and peacefully among us.'

But if Goethe's idea was correct, the glaciers of the Alps and the Scandinavian mountains must in earlier times have been much larger than they are at present. Goethe fully realized this. In November 1829 he set down the following comment in his notes: 'For all that ice we need cold. I would wager that an epoch of great cold passed over Europe, at any rate.'

While von Hoff still believed that boulders had been transported by drifting ice floes (a view which eventually became established and for decades was regarded as ultimate truth), Goethe had already had the brilliant vision of an Ice Age. In the last year of his life he reverted to this thought: 'If a great cold spell spanned a large part of northern Germany with a single surface of ice, we may conceive what destruction the chaotic ice floes would wreak when the great mass thawed, and how it would be bound to carry granite blocks farther south.'

The contention has been made that Goethe was borrowing the theories of another man. But Goethe never knew or heard of this predecessor: the engineer Venetz, from the Swiss canton of the Valais. From 1821 on, Venetz had been talking and lecturing untiringly in the effort to persuade the naturalists of his native land that ice had been a force of incalculable power in prehistoric times. During Goethe's lifetime people only shook their heads over his wild notions; none of his ideas reached Weimar.

By the time Venetz finally published a treatise on his investigations, and thereby won a few prominent allies in the scientific world, the great German poet had been resting in his grave for a good year.

In the Glaciers of the Alps

Originally Venetz had been concerned with glaciers for a thoroughly humdrum professional reason. He had been assigned by the Swiss government to study the great rivers of ice and suggest ways to avert the dangers of avalanches. In the course of his travels and surveys, however, the idea gradually developed in his mind that the glaciers could have been considerably larger at some time in the past than they were at present. He tried to find evidence for this. At the eighth annual meeting of the society of Swiss naturalists held in 1821 he read a paper which argued that the presence of moraines proved there had been an enormous extension of the glaciers at some prehistoric period. Eight years later, in 1829, he no longer restricted himself to the glaciers and moraines of Canton Valais; he had expanded his theory to include all of prehistoric Europe. Mighty glaciers, he asserted, had once covered a large part of Europe; it was because of them that erratic blocks had been found not only in the Alpine region, but also in northern Germany.

Here, then, was the same idea that Goethe had tentatively expressed. It became almost an obsession with Venetz, for the Valaisan engineer was a stubborn man. When he aroused no interest at congresses of naturalists, he went to Lausanne and called on Jean de Charpentier, professor of geology and director of the salt mines of Canton Vaud. Charpentier came from a Huguenot family of scholars that had already produced a number of great geologists and scientific experts on mining; in Switzerland he was regarded as the supreme authority on all phases of the earth sciences. He was the right man for Engineer Venetz.

As Jean de Charpentier sat talking with Venetz, he suddenly recalled a conversation he had had fourteen years before with an old chamois hunter named Perraudin. During a mountain expedition, he had spent a night in Perraudin's hut and talked with him about the glaciers in the vicinity of the St Bernard Pass. Perraudin was of the opinion that the glaciers must once have covered far more territory than they did at present, as could be seen from the boulders near the town of Martigny. Those boulders were too large to have been moved by water, the chamois hunter had reasoned, and therefore could only have been carried by the ice from the heights of the St Bernard to Martigny.

'Although good Perraudin confined the travels of his glacier to Martigny, probably because he himself had scarcely ever gone any farther,' Charpentier later related, 'I thought this hypothesis so strange and extravagant as not to be worth considering.' But now Venetz, a man trained in science, came along and asserted that gigantic glaciers had once covered all of Switzerland and large parts of the rest of Europe. That sounded a good deal wilder than the fancy of the old chamois hunter.

'At first,' Charpentier wrote, 'the idea struck me as altogether foolish. It seemed to stand in stark contradiction to all the principles of physics and geology.' The geologists had discovered that the climate of Europe in earlier times (the Tertiary, as we now know) had been considerably warmer than at present. Elephants, rhinoceroses, and other tropical animals had lived in the Europe of the distant past. 'How was it possible to imagine a region in which palms once throve to have been covered by a glacier?'

But Charpentier was as thorough a man as Venetz was a stubborn one. In order to convince the engineer of the wrongness of his views, the Lausanne professor of geology began studying erratic deposits in Switzerland. To his own surprise he discovered that Venetz could not be refuted; on the contrary, everything bore him out. In 1834 Charpentier delivered a lecture in Lucerne in which he virtually repeated Venetz's arguments. And since Charpentier was not an obscure engineer, but a renowned scientist, his arguments caused a sensation far beyond the borders of Switzerland. They brought about a break-through and served as a starting point for unravelling the riddles of the Ice Age.

The Ice Age – a Catastrophe?

Two years later a committee of prominent scientists went into the Alps to study the glaciers and moraines in the light of Charpentier's theories. Among the group were the German botanist Karl Schimper, a young zoologist, Karl Vogt, who was later to become one of the stoutest champions of Darwinism, and the Swiss zoologist and palaeontologist Louis Agassiz, who at the time had not quite reached his thirtieth year, but was already counted among the foremost scientists of his age. If a man like Agassiz confirmed Jean de Charpentier's conclusions, the victory of the new ice theory over the old deluge theories would be assured.

Louis Agassiz came from a family of Calvinist ministers. As a student he began specializing in the study of fossil fish, corresponded with Cuvier and Alexander von Humboldt, and despite his youth was already being called a second Cuvier by many specialists. In 1832 he was appointed professor of natural history at the then still inconsequential university of Neuchâtel. He proceeded to turn the small cantonal capital into a centre of scientific research. When Agassiz joined forces with Charpentier and in a succession of reports enthusiastically supported the glacier theory, the entire scientific world paid heed.

Charpentier and Agassiz did not entirely agree on matters of detail. Charpentier thought in terms of periodic raisings and sinkings of the Alps; he maintained that the Ice Age had been a purely local phenomenon of the alpine regions. Agassiz, on the other hand, adhered to the catastrophe theory of his great master, Cuvier. The Ice Age, he said, had been brought on by a sudden calamitous cooling, and had embraced wide areas of the earth. Europe, northern Asia, and North America had become one gigantic sheet of ice. He cited the mammoths in the Siberian ice as instances of that world-wide catastrophe; they had been caught by the sudden outbreak of cold and thus locked into the ice before they could decay.

Agassiz was right in presuming that the Ice Age had once affected the entire northern third of the earth. He was also right in attributing moraines not to the work of drifting ice floes, but to the activity of huge glaciers. But because of his loyalty to the old cataclysm theory he in effect only replaced the Biblical Deluge with an equally violent and catastrophic freeze-up.

When Agassiz published his Ice Age theory, he lost a good

friend. Karl Schimper, the botanist whom he had known since his student days in Munich, and with whom he had explored the alpine landscape, could not accept Agassiz's view; there had been several periods of cold and warmth; erratic blocks, he held, had not been transported by glaciers, but by floating icebergs. Schimper had a weakness for setting forth his ideas in verse. In 1837, in a poem, he invented the word *Eiszeit* – Ice Age.

Agassiz quoted Schimper now and then in his lectures, and picked up the phrase 'Ice Age'. Schimper took this as a violation of his rights of priority, and waxed indignant. Subsequently, he developed into an unbearable grumbler and nuisance, constantly bombarding Swiss naturalists with circular letters charging that Agassiz had stolen the word 'Ice Age' from him, and the entire theory as well. Finally he became a drinker, and died insane.

Schimper's drifting icebergs, however, won out over Louis Agassiz's glaciers – for a while at least. In England, meanwhile, a geologist had examined the phenomenon of the Ice Age along the lines of Karl Ernst von Hoff. He would have it that the Arctic Ocean with its boulder-laden ice floes had once washed Europe's coast; this seemed far more likely than the hypothesis of a huge and violent glaciation of broad areas of the globe. The man who took this view was Charles Lyell, the great renewer of geology.

A curious situation thus arose. The adherents of the old, mistaken catastrophe theory were also, following the example of Louis Agassiz, the most fervent supporters of the correct glacial theory. The adherents of the new, correct evolutionary theory, following the example of Charles Lyell, were simultaneously the strongest advocates of the erroneous drift theory.

The Renewer of Geology

A revolutionary in science does not have an easy time putting his new ideas across when he is dependent on his teaching position, his sovereign, the security of his livelihood. For that very reason the great innovators in the history of the sciences have so frequently been people of independent means who could follow their own bent.

Charles Lyell is a typical example of such a type. He was the eldest son of a rich landowner, had the best of educations, and while at Oxford could pursue all the subjects that interested him, from law to zoology. He had the means to travel over half of Europe while still a young man; and later he spent a large part of his life touring the world. For a time he gave lectures at King's College in London; but when offered a professorship, he declined in order not to be tied down. He lived solely for his scientific interests and avocations, and was probably one of the happiest men of his age.

Lyell had only one handicap to combat: he had a serious defect of vision. An ardent geologist, naturalist, and traveller needs good eyes. Lyell overcame this obstacle by marrying Mary Horner, the daughter of the English geologist and explorer of Indonesia, Leonard Horner. Mary became his constant assistant and companion; through her sharp eyes he henceforth saw the earth and its phenomena.

Even when in due course he inherited the family estate and the title of baronet, Lyell did not have to shoulder the burden of land management. He had already so distinguished himself that his family did not dream of asking him to bury himself in the country. Like Alexander von Humboldt and his later friend Charles Darwin, Lyell remained all his life a free-wheeling scholar.

Originally, Lyell had intended to enter law, and had studied entomology only as a hobby. But at the age of twenty-one his interest in geology suddenly awakened. A trip with his parents through France, Italy, and Switzerland seems to have stimulated him to intensive geological studies. Only a year later he had already become a member of the Geological Society of London. On other travels he met Cuvier and Alexander von Humboldt in Paris. In that city he also met a person whose views were to exert a decisive influence upon him. This was Constant Prévost; he was professor of mineralogy and geology at the Paris Athenaeum. Prévost had criticized the Cuvier catastrophe theory on more than one occasion, and had expressed ideas similar to those of von Hoff: that geological changes in earlier ages did not differ essentially from the events currently taking place on and inside the earth. But since Prévost was a shy man who did not want to wrangle with Cuvier, he gave only the most timid utterance to his opinions.

He struck up a friendship with Lyell, and accompanied him to England, where the two went on a number of expeditions. Lyell found his friend's ideas worth serious attention. And since the Englishman, though ten years younger than his French colleague, had none of Prévost's timidity, he at once set about assembling all possible proofs of slow changes in the earth. By 1827 Lyell had completed the greater part of his *Principles of Geology*, the great work of his life, which was to become the foundation for a new age of natural science, the age of actualism and evolution, of Darwin and his fellow warriors.

The Burial of Diluvianism

Among Lyell's teachers at Oxford had been the orthodox Diluvian William Buckland; he lectured on both geology and theology, distinctly favouring theology. Buckland clung tenaciously to the Deluge idea. In his chief work, *Reliquiae Diluvianae*, he declared that he could find only 'a single possible explanation' for the existence of fossil animal remains and human bones. That would have to be 'the force of a short-term flood'. Most of the bones of the animals in English caves, he asserted, belonged to the same species as the bones in the deposits and rock clefts of the Continent. This amazing coincidence of fossils was clear proof of a universal Deluge which must have taken place only a few thousand years before, happily corroborating the account in Genesis.

William Buckland was the last serious advocate of the Deluge idea. The theory of the great Flood died with him, after having dominated the thoughts of geologists and palaeontologists for a century and a half. For Buckland himself that theory had long since ceased to bear any relationship to science; it had become a profession of faith. But by the time Buckland had at last become Dean of Westminster, he had really ceased to believe that his geological Deluge was one and the same as the Biblical Deluge. He had been robbed of his faith by none other than Agassiz.

At a naturalists' congress in Glasgow, Agassiz had taken occasion to begin a disputation with Buckland. It was somewhat nonplussing when Agassiz countered the words of Holy Scripture with descriptions of ground moraines and end moraines, glacier polishing, erratic blocks, and other geological phenomena. A good geologist in spite of his pious belief in the Flood, Buckland found himself unable to answer the Ice Age expert from Switzerland. His uncertainty increased when he learned that Agassiz was by no means one of those wicked atheists determined to explain all natural events materialistically, but as godly a man as himself, from a good Calvinist family of theologians.

The last of the Diluvians therefore resolved upon the difficult step of following Agassiz and replacing the Deluge catastrophe by a glaciation catastrophe. In so doing he tried to save what he could of the Deluge idea. Perhaps the ocean had flooded the dry land and then frozen, as the Arctic seas are still frozen. But, as Buckland at last realized, this violent event could scarcely have had anything to do with the Biblical Deluge. The glaciation had been 'the last of the great geological revolutions'; it had taken place 'in a period preceding the creation of man', and therefore could not be equated with the relatively mild Biblical flood.

Thus the last serious spokesman for Diluvianism buried the theory. It lives on only in those visionary cosmogonies which would have the moon or asteroids plunging into our earth, thereby producing catastrophic floods.

When Lyell wrote his *Principles*, therefore, he no longer needed to oppose the Biblical account. On other fundamental geological questions, however, he had to declare war on his teacher, Buckland. The central problem had remained the same as in Cuvier's day: Had Mother Earth acquired her present appearance as the result of catastrophic upheavals or slow changes? This was von Hoff's and Goethe's question too. Inextricably linked with it were the great questions of the evolution of life and the origin of man.

Slow Evolution?

Anyone who wishes to construct a new system must first demolish the outmoded structure. Lyell did so in his *Principles*, with a thoroughness even more impressive than that of von Hoff. The introductory chapters were directed at those of his contemporaries who believed in the Bible. He demonstrated that the greatest obstacle to proper study of the earth's history had hitherto been theological prejudice, above all obstinate adherence to the data of the Creation story and to Biblical chronology. Then he proceeded to dispel the catastrophe theory. He described the different ages of the earth, their climatic conditions, and the changing distribution of water and land; he cited innumerable examples validating his view that there had never been violent revolutions in the history of the earth, but only step-by-step transformations whose effects were cumulative.

Small causes, according to Lyell, had worked even the greatest of changes upon the surface of the earth in the course of inconceivably long spans of time. From observation of these changes geologists had concluded that there must have been upheavals, but this was an error of perspective. The history of the human race would also, if viewed thus speeded up, look like a succession of catastrophes and revolutions. Lyell constantly returned to the factor of 'time'. He made it the core of his new geology.

The *Principles of Geology* was published from 1830 to 1833. Its success was amazing; this dense, three-volume book, dealing with so rigorous a subject as geology, went through one edition after the other. The sixth was published by 1840; the twelfth shortly before Lyell's death in 1875. Each time Lyell expanded and improved his text. In the course of time he had to alter fundamentally some of his views and conclusions; and it is to his credit that he did so at once, as soon as he realized that he had been mistaken.

One of his major errors was his attitude towards evolution. From his theory of slow changes in the surface of the earth, logic should have called for the idea that plants and animals had also evolved slowly. But up to the ninth edition of the *Principles* Lyell clung to Cuvier's dogma of the immutability of species. He had read Lamarck's writings but could not fit them into his philosophy. The only concession he would make to Lamarckism was to grant an extremely minor adjustment of organisms to their environment. Fossils at first interested him only as evidence for his geological theory. He observed that certain types of marine animals recurred in all epochs, and took this to mean that the living conditions of these species could not have been much different in earlier times than they were at the present time.

Lyell became a passionate advocate of evolution only after Darwin made his appearance. From then to the end of his life he was a close friend of Darwin's and one of the most successful of the pioneers of Darwinism. Under the influence of Darwin he rewrote important sections of his *Principles of Geology*. In fact, eight years *before* the publication of Darwin's book, *The Descent of Man*, Lyell went on record for the idea of man's descent from animals.

Although actualism had a number of weak spots, which today are rightly criticized, geologists have on the whole accepted Lyell's principle of seeking the key to geological processes of the past in those that can be observed in the present. With Lyell there began the era of dynamic geology.

Lyell's authority was so great that his contemporaries failed to notice his errors. Wherever he could, he corrected these himself; but some things he could not correct because research had not made sufficient progress in certain fields. Among these was his attitude towards the concept of the Ice Age. He rejected the glacier theory of Louis Agassiz, on grounds that it smacked too much of violence; instead he preferred the drift theory.

A whole generation of progressive geologists and evolutionary biologists were partial to the drift theory. The first Arctic explorers were partly responsible for this. Men like Parry, Ross, and Scoresby had painstakingly described the calving of Greenland glaciers, and the drifting and stranding of icebergs and ice floes. It seemed logical to conclude that similar phenomena had characterized the Ice Age. On the strength of such accounts, Darwin, too, believed unswervingly in the drift theory until very late in life; the thought of a gigantic glaciation spread over vast parts of Europe, northern Asia, and North America struck him as the wildest kind of speculation.

Part of the hostility between the adherents and the opponents of the theory of evolution may have arisen out of this difference over the nature of the Ice Age. Agassiz was one of the sharpest critics of Lyell and Darwin; hence Lyell, Darwin, and the other evolutionists treated his glaciation theory with the utmost scepticism. In addition, Agassiz had linked it with the old, damnable doctrine of catastrophes. In this form it was all the more unacceptable to followers of Lyell and Darwin.

The Ice Age question was not finally settled until 1875. The two great antagonists, Lyell and Agassiz, did not live to see its resolution.

Scratches in the Limestone

On 3 November 1875 a meeting of the German Geological Society took place in Berlin. One of the most noted of Sweden's geologists, Otto Martin Torell, rose to deliver a paper. He discussed some excursions he had undertaken with two German colleagues to the Rüdersdorf limestone hills near Berlin. He had barely finished when a vehement, in fact tumultuous, discussion ensued. Rightly, Torell's lecture was taken to be a major scientific sensation.

Torell had spent some time living on the inland ice sheets of Greenland and Spitzbergen. He was familiar with the phenomena of glaciation. Contrary to the prevailing view, he believed that all of Scandinavia had once been entirely covered by glaciers. In a virtually unknown treatise by his fellow countryman Nils Gabriel Sefström, written in 1838, Torell had found an account of the Rüdersdorf limestone hills. According to Sefström the limestone there showed curious scratches, scars, and polished surfaces which could only be explained as the effects of ice. This statement suggested to Torell that the observations he had made in Scandinavia must also apply to Central Europe. The scratches and polished surfaces that Sefström had seen in Rüdersdorf might possibly provide the proof, he thought.

The excursions to Rüdersdorf had yielded splendid confirmations of the theory. Torell had brought back stones marked by clear parallel scratches which he now showed to the assembly. The scratches, he explained to the aroused geologists, unmistakably pointed to the movement of a glacier. Such total similarities existed between the diluvial formations of Scandinavia and of the North German plain that there was nothing for it but to assume the same origin for both phenomena.

Thus, Torell continued, not the ice floes of the Arctic Ocean but the glaciers of Scandinavia had covered Europe during the Ice Age. Moreover, the alpine glaciers had moved up from the south, as Agassiz had supposed. Things must have been similar in North America and northern Asia: glaciers from the high mountains had not advanced with catastrophic speed, as Agassiz thought, but by slow, steady, almost imperceptible stages had gradually covered a large portion of the land. Then the glacial fronts receded just as slowly, steadily, and imperceptibly. How often had this process taken place in that epoch which, on the basis of the erstwhile belief in the Deluge, was still called the *Diluvium*? (It is now called the Pleistocene – a gesture towards Lyell, who in 1830 coined this word [from Gr. *pleistos*, most, and *kainos*, recent].) Once? Several times? At his lecture in Berlin, Torell did not yet venture to make any definite assertions.

At that since celebrated Berlin session, the geologists continued for a while to argue over the scars on the Rüdersdorf limestone. But Torell had presented such a telling thesis that the German geologists at once set to work examining the legacies of the Ice Age on the North German plain. Within four years a young man from Leipzig, then barely twenty-one, delivered the deathblow to the drift theory.

Cold Periods and Warm Periods

Albrecht Penck was the name of this talented young man. He travelled through North Germany and Scandinavia, and found many confirmations of Torell's views. Subsequently, he studied moraines and traces of glaciation in the Alps. His book on the glaciation of Germany, the first scientifically sound description of the Ice Age, won a prize which provided him with funds to study the effects of the Ice Age in Scotland, the Pyrenees and other countries and mountain ranges. Penck was also the first geologist who incisively proved that the European Ice Age had in reality consisted of several periods of glaciation with intervening periods of greater warmth.

There have been four major glaciations in the Pleistocene of the Old and New Worlds. In Europe the periods are named after small rivers in South Germany marking the lines of advance of the glacier: the Günz-Mindel, Riss, and Würm Ice Ages. In

North America they are named the Nebraskan, Kansan, Illinoian, and Wisconsin Ice Ages, after those states where the deposits of the glacier are most clearly to be seen. The intervening interglacial ages are called in America Afronian, Yarmouth, and Sangamon ages, in Europe Günz-Mindel, Mindel-Riss, and Riss-Würm warm ages. During these times there prevailed in both the Old and the New World a mild, warm climate, in places subtropical.

This division of the Pleistocene is by now widely known; but it is not very old at all. It was devised by the German palaeontologist Wolfgang Soergel between 1910 and 1925. Soergel thoroughly studied the fossils of the Pleistocene. He determined which species of animals had lived in the various glacial ages and which in the intervening warm ages. The fossil remains of plants were an excellent clue to the vegetation and hence the climate in the various warm and cold sub-ages.

The usual picture of the Ice Age, familiar to most of us, is based on Soergel's remarkable essay, *Subdivision and Absolute Chronology of the Ice Age*. The Pleistocene Ice Age began six hundred thousand years ago, he concluded. The four advances of cold and the three warm interglacial ages were spread out over that span of time. About twenty-five to twelve thousand years ago, according to region, the Pleistocene could be considered to have ended.

But it has since been discovered that Soergel's Ice Age pattern does not suffice. In North Germany only three glaciations have been discerned, now named the Elster, Saale, and Vistula Ages. But in South Germany there appear to have been not only the four usually reckoned, but a fifth: the Danube Ice Age. Moreover it turned out that the periods of glaciation were not evenly cold. The glaciers advanced and their fronts receded constantly within any given glacial age. Probably every Ice Age actually consisted of at least two advances of cold and an intervening interval of warmth. During the last glaciation, the Wisconsin Ice Age, three such advances of cold have been identified, one of them characterized by unusually low temperatures and extraordinary dryness.

The Deep Sea and the Earth's Past

The standard dating has also been considerably shaken in recent times. German oceanographers had already pointed out before the last war that the depths of the ocean, with its thick sediments, could give us some information on Pleistocene climatic alternations. In 1949 the head of the geology department of Columbia University, Maurice Ewing, decided to look into this idea. He sent two of his associates, David B. Ericson and Gösta Wollin, on an oceanographic expedition to obtain a new and better dating of the Ice Age.

On further expeditions stretching over almost fifteen years, the two scientists systematically explored the ocean bottom, collecting the sediments and studying the innumerable fossil microorganisms which had been deposited in the ooze. The number of layers alone was cause for wonder. And, as had been predicted, the formations at the bottom of the ocean deeps furnished distinct traces of past climatic conditions. In their book *The Deep and the Past* (1964) Ericson and Wollin drew up a new chronological table for geology.

According to their reckoning, the Pleistocene had lasted twice as long as had previously been assumed. It began about one and a half million years ago. Above all, according to Ericson and Wollin, the interglacial ages had occupied much longer periods of time than had hitherto been suspected. The proofs they offered for their new calculations sound extremely convincing. Whether their dating of the Ice Age is generally valid for the whole world, research should determine in the near future.

Ericson and Wollin's new dating could be extraordinarily important to evolutionary scientists. For in the Pleistocene in particular an astonishingly large number of species flourished, changed, and vanished with apparent suddenness in what seemed to be an extremely short span of time. Our own species, too, according to the old dating had only a relatively brief span in which to develop from primitive ape-man to culture-creating and art-producing *Homo sapiens*. If the time span is doubled, the emergence of prehistoric and early man appears far more reasonable.

The southern as well as the northern hemisphere was affected by the Pleistocene Ice Age. In the south, the glaciers spread out from Antarctica. The Antarctic Ocean froze as far as the coasts of Australia and the southern portion of South America. Argentina, Chile, and New Zealand became glaciated, or frozen, plains.

So much water was withdrawn from the ocean by the formation of these tremendous masses of ice that the level of the sea presumably dropped three hundred feet or more. Mammoths grazed on the Dogger Bank in the present North Sea. It was possible to walk on dry land from Europe to England and North Africa, from north-eastern Asia to North America, from India to Indonesia, and from Australia to Tasmania.

That is, almost the entire globe was directly or indirectly affected by the Ice Age.

But what had caused such vast expansions of the glaciers? Why had the annual average temperature dropped between six and eight degrees centigrade? What earthly or cosmic events were to blame for the fact that rivers and sheets of ice turned so large a portion of the earth's surface into cold, barren tundra?

The Many Riddles of the Ice Age

There are certain themes in nature and in the history of thought that perennially appeal to the imagination of men. Among these are the legend of the Deluge, the myth of Atlantis, the possibility of contact with supernatural realms. The cause of the Ice Age is one such theme. Whole libraries have been written about it. Interesting theories and striking deductions alternate with preposterous fables of volcanoes spewing ice and stars radiating cold to bring about the requisite lower temperatures.

Wilhelm Bölsche gave an apt description of this sort of thing in 1920: 'Anyone who has had occasion to publish some findings about the Ice Age finds to his horror that innumerable manuscripts in terrifyingly bulky parcels collect on his desk, with and without return postage, whose senders invariably proclaim: I too have found a solution to the riddle of the Ice Age!'

Perhaps it was to discourage such amateurs that a leading German geologist towards the end of the nineteenth century was wont to cry at the top of his voice: 'We do not know the causes of the Ice Age!'

Would that be true today? As a matter of fact, Ice Age problems have not been made clearer by the advances of science, but if anything have become more complex. Geologists have determined that there was not just a 'Pleistocene' Ice Age. Five hundred million years ago, in the Precambrian and at the beginning of the Cambrian, great blankets of cold descended upon eastern Asia and the South Pacific region. And again 240 million years ago, in the Permian period, when many species of primitive saurians already inhabited the inland seas, deserts, and fern forests, an unusually harsh and long-lasting Ice Age descended upon India, Australia, South Africa, and South America. In addition, other Ice Ages or at least local glaciations are suspected in the very ancient Algonkian, with others occurring about 350 million years ago in the Silurian, and sixty or seventy million years ago towards the end of the Cretaceous.

Wandering Poles, Luxuriant Forests

Serious theories about the possible causes of the Ice Age must attempt to deal with this curious periodicity. For a long time the hypothesis of the wandering poles was favoured. It received fresh support when Alfred Wegener, the German geographer, propounded his continental drift theory. According to Wegener, the continents drift like ice floes on the viscous layers of the earth's interior. The result is that at different times a different continent is near the North or the South Poles. When this happens it experiences an Ice Age, as happened in the Precambrian of China and Australia, in the Permian to a large part of the southern hemisphere, and in the Pleistocene to Europe and North America. Due to the continental drift, according to Wegener, Antarctica was free of ice a quarter of a billion years ago, and tropical plants flourished in Greenland and Spitzbergen sixty million years ago.

This idea of drifting continents likewise encourages the imagination to all manner of speculation. Wegener's theory can serve to explain all imaginable climatic conditions. We need only postulate the position of the continents *vis-à-vis* the Poles in earlier times in such a way that they correspond with the climates of these epochs, and all problems of cold and warm ages seem to answer themselves. At one time the basalt floor of the oceans was thought to be much too rigid to permit the continents to drift like ice floes. Nevertheless, the last word for or against the theory has not yet been spoken.

This is also true of the second hypothesis, which has excited equal discussion and which also has much to commend it, much to take exception to. It was first put forward by Svante Arrhenius, the Swedish chemist (1859–1927). According to Arrhenius, the climate on our planet depends on the carbon dioxide content of the air. Air saturated with carbon dioxide produces an effect like that of the windows in a greenhouse: it retains the warmth of the sun's rays. In times of intensified volcanic eruptions much carbon dioxide gas may be poured into the atmosphere, in which case the temperature of the earth rises; the globe undergoes a warm period. If, on the other hand, an unusually large amount of carbon dioxide is withdrawn from the atmosphere by luxuriant vegetation, the temperature drops and an Ice Age follows.

This line of argument seemed particularly persuasive because it agreed with the geological and palaeontological facts. The time of intense volcanic activity in the Palaeozoic era was followed by the period with the most luxuriant vegetation of all time: the Carboniferous. The marsh and rain forests of the period withdrew enormous quantities of carbon dioxide from the atmosphere and fixed it in the form we find it today: as coal. According to Arrhenius, the consequence was an Ice Age in the Permian which wiped out many of the species of plants that had flourished in the Carboniferous.

A similar process presumably took place once again in more recent times. The volcanic eruptions of the Mesozoic era again saturated the atmosphere with carbon dioxide. There followed, in the Tertiary, an unusual increase of vegetation – which became our present lignite or brown coal. But once more a carbon dioxide deficiency ensued, followed by the more recent Ice Age, the Pleistocene.

A number of weak points have been detected in this theory of Svante Arrhenius. Above all it does not explain the periodicity within the Ice Ages, the rhythmic alternation of cold and warm periods which have been observed in the Permian as well as in the Pleistocene Ice Ages. Consequently, many theorists have turned from earth-centred hypotheses and looked for cosmic causes.

Earth's Eccentric Orbit

As early as 1837 the French mathematician Poisson theorized that our earth passed through sometimes warm and sometimes cold regions of space; this, he said, would explain the major climatic changes in the earth's history. Eventually astronomers were indeed able to demonstrate that two large dark clouds composed of countless tiny particles of matter exist in the part of space through which the sun passes with its family of planets.

Every 220 to 250 million years the sun enters one of these dark clouds. Its radiation of light and heat is slightly absorbed by the particles. This might produce a temperature drop on the planets of the solar system.

The increase and diminution of sunspots has likewise been suggested as an explanation for the Ice Ages. The more sunspots the sun has, the less the intensity of its radiation. In 1893 the Dutch scientist Eugen Dubois, the discoverer of the Java ape-man, offered an ingenious theory of sunspots. The spots, he maintained, are plain signs that the sun is gradually cooling. As yet there are periodic alternations of temperature; the number of sunspots increases and diminishes in a regular cycle. Any unusual increase in these spots, Dubois suggested, would be bound to bring about an Ice Age on earth.

All these attempts at explanation, interesting and cogent though they seem, cannot compete with an extremely complicated theory based on the eccentricity of the earth's orbit. That orbit is no more a perfect circle than the earth itself is a perfect sphere. Our planet moves around the sun in an orbit which sometimes has nearly the shape of a circle, sometimes approaches an elongated ellipse. The sun does not lie precisely in the centre of this ellipse. Normally, then, the earth periodically has times when it is nearer and times when it is farther from the sun.

In addition, the inclination of the earth's axis, which at present amounts to $23\frac{1}{2}$ degrees, is subject to variations. There are a number of other astronomical peculiarities of the orbit, each of which might cause temperature deviations, though they would be minor ones. Nevertheless, when all the factors coincide, that is, when extreme eccentricity of the earth's orbit occurs at the same time as maximum tilt of the axis, the combined effect may so reduce the temperature on our earth that an Ice Age results.

In 1842 Alphonse Joseph Adhémar, who taught mathematics in Paris, wrote an enthralling book based on these facts; his contemporaries read it with much the same passion they would feel for the novels of Jules Verne twenty years later. Adhémar postulated that an Ice Age engulfed the northern and the southern hemispheres alternately every twenty-one thousand years. He was, to be sure, an adherent of the catastrophe theory. If the polar ice increased beyond a certain critical point, he reasoned, the earth's centre of gravity would shift and the sea would pour over the continents. Thus the Ice Age passed into a universal Deluge.

These fantastic ideas fell into oblivion. But attempts to explain the phenomena of the Ice Ages by the eccentricity of the earth's orbit did not cease. They culminated in the researches of Milutin Milankovich, the Yugoslav astronomer, who, beginning in 1930, spent more than a decade working on the problem. From the periodic changes in the elements of the orbit, the obliquity of the ecliptic, the procession of the equinoxes and other astronomical factors, Milankovich calculated the intensity of solar radiation for a great many different latitudes of our earth during the last six hundred thousand years. The radiation curves he obtained corresponded amazingly with the geologists' dates of Ice Ages.

Nine times in the last six hundred millennia, according to Milankovich's calculations, the coincidence of periodic astronomical events so diminished the intensity of solar radiation that an Ice Age necessarily resulted. The nine dates he obtained corresponded very well with the nine previously determined onsets of cold during the four glaciation periods of the Pleistocene.

All this sounds highly convincing. However, Milankovich's radiation curves for older geological epochs have yielded rather uncertain results. They can scarcely be synchronized even with the new Ice Age chronology of Ericson and Wollin. Nevertheless, for lack of any better theory Milankovich's approach still seems to offer the most likely explanation for the events of the Ice Ages.

World Ice and Falling Moon

Imaginative outsiders still go on spinning their theories. Hanns Hörbiger, the Austrian engineer, maintains that the Ice Ages were caused by the earth's passing through a zone of ice particles which drift through space. Others have invoked a wide variety of comets, meteors, planetoids, or moons plunging down upon earth, thus inducing a sudden shift in the earth's axis, or brusquely knocking the earth out of her orbit, or even bringing her briefly to a stop. The result of such violent interference with the earth would be a catastrophic deluge, a rain of mud and nitrogen poisoning, which finally ended in glaciation.

All of these theories betray the deep hold of the old cataclysmic notions on the minds of many people. The imagination, it would seem, prefers the sudden catastrophe to the concept of accumulated slow changes.

The Siberian mammoths appear to be a favourite proof of such violence in nature. Siberia never was, as a modern catastrophe theorist has flamboyantly said, 'a tremendous ice chest filled with well-preserved mammoth cadavers'; but it is true that a number of remarkably well-preserved mammoth corpses have been found in the Siberian ice. And scientists as well as laymen have seriously wondered whether these huge beasts could have died a natural death or whether they were sudden victims of violent change.

The stomachs of some of the huge animals have disclosed lumps of undigested or half-digested plants, tundra grasses, and coniferous shoots. In some cases their flesh was found to be so fresh after thawing that men as well as predatory animals and dogs could eat it. Participants in a scientific congress at the Geological Institute of St Petersburg were once actually served mammoth steaks; they came from a mammoth calf that had been frozen into a block of ice and so transported to the Russian capital as if in a freezer.

The mortality of so many mammoths in Siberia does present palaeontologists with a number of mysteries. But the true story of this typical Ice Age mammal is far more interesting than all the visionary notions about world ice, falling moons, and similar cosmic accidents.

The Beasts in the Freezer Locker

It had been known for centuries that there were mammoths in the Siberian ice. Tunguses and other hunters of the tundra came on the huge cadavers of these frozen animals again and again. They would usually wait for the blocks of ice to thaw in order to obtain the tusks, and would then feed the meat to their dogs, if it were not first taken by hungry bears, wolves, and foxes.

Gradually a flourishing business developed out of the trade in tusks. Many thousands of pounds of mammoth ivory were collected year after year, especially in northeastern Siberia, between the Lena and Indigirka rivers, and in the New Siberian islands to the north. A third of the ivory that has entered commerce up to modern times came from mammoths.

The ivory dealers gradually realized that the mammoth was no relation to unicorns, behemoths, or other fabulous beasts. They had only to examine the tusks closely to see that they were elephant teeth. But how could elephants ever have lived in cold Siberia?

The subject was hotly discussed at the Petersburg Academy; two full centuries before Wegener, Russian naturalists toyed with the idea that the Pole might have shifted. Siberian elephants, they concluded, were identical with the present-day elephants of southern Asia. Hence, where the Arctic tundra with its permafrost now stretched for endless miles, there had been in the past a warm, in fact tropical, climate. The shift of the Pole changed the climate of Siberia and thus condemned the elephants to death by freezing.

Outside the Russian borders, little was heard of these reflections and researches of Russian academicians. Reports on the Siberian mammoth reached the West only rarely, and scantily, until the beginning of the nineteenth century. Finds of mammoth bones in Europe were another matter and western scientists were much preoccupied with these. In 1799 Johann Friedrich Blumenbach of Göttingen, the founder of zoology in Germany, became aware that the mammoths belonged to a species different from the two species of living elephants. He therefore gave it the name *Elephas primigenius*, 'first-born elephant'.

Blumenbach could not know that the mammoth was by no means the 'first' elephant, but a distinctly late descendant of a proboscidian order that once boasted many species and was distributed over the whole world. Nor could he suspect that his 'first-born elephant' was identical with or closely related to the mysterious Siberian mammoth. But as it happened, in the very year of 1799 in which Blumenbach found a scientific name for the European mammoths, a Tungus named Ossip Shumakhov took a step that led to the solution of the mammoth enigma.

Shumakhov had come upon a completely preserved mammoth in the ice at the mouth of the Lena River. The huge cadaver

Mammoth attacking – drawing scratched on ivory, from the
La Madeleine cave.

was inaccessible, for it was frozen into an enormous block of ice.
Among the Tungus tribes at the time there was a superstition
that mammoths were monsters that lived underground, rather
like gigantic moles, and perished the moment they saw the light.
Hence – the reasoning was clear to any Tungus – anyone who
found a mammoth cadaver was doomed to die, along with his
entire family.

Consequently, two impulses contended with Shumakhov. He
wanted to take the tusks when the block of ice thawed, so that
he could sell them to Boltunov, the Russian ivory dealer who
traded in the vicinity; but at the same time he feared to risk
death. In the next two years he visited the block of ice several
times. The thawing process had just begun; it was still not
possible to get at the ivory. After a while his superstitious fear
increased to such a psychotic point that he actually fell seriously
ill.

The illness did not last long, however. And when it departed,
Shumakhov's superstition went with it, yielding to a healthy
acquisitiveness. He began to disbelieve the legends of his tribe
and in 1803, four years after he had discovered the mammoth, he
decided to lead the trader Boltunov to the block of ice. Meanwhile
the ivory had thawed out of it. The animal's head was uncovered
and already beginning to decay.

Shumakhov sold the tusks to the Russian for fifty roubles.
Boltunov, however, did not content himself with the mere pur-

chase; he took a good long look at the thawing cadaver and even made a drawing of the animal. It was not a very good drawing, and certainly not a correct one, for the animal was no longer intact. Hence the head in Boltunov's drawing looks like a gigantic pig's head; the eyes are placed where in reality the openings to the auditory passages should be. But the body was rendered quite correctly.

The ivory trader took a step further. He sent his drawing, together with a description, to the Petersburg Academy. There it fell into the hands of the English zoologist Henry Adams. Adams passed the drawing on to Göttingen, asking his German colleague Blumenbach for an opinion. Blumenbach at once decided that this must be his *Elephas primigenius*. Cuvier, too, saw the drawing and concurred with Blumenbach.

Adams meanwhile had not been idle. By 1806 he had organized an expedition bent on rescuing whatever parts of the mammoth had not decayed or been eaten by bears, wolves, wolverines, and foxes. He went by reindeer sleds to the mouth of the Lena River and sought out Shumakhov. By this time the native realized fully that his mammoth was far from a harbinger of bad luck; on the contrary, it was obviously going to mean the best of luck to him. For Professor Adams paid him generously for his services as a guide.

Three quarters of the skin, one ear, and the entire skeleton with the exception of one forefoot was still left of Shumakhov's mammoth. The skin was covered with thick, woolly hair. Adams had great difficulty in packing the remains of the huge beast on his sleds. Ten men were scarcely able to move the hide alone, after it had been stripped from the carcass. On the long trip to St Petersburg 'Adams' mammoth', as the find was called, was knocked about so much that soon not a hair was to be seen on the skin. Nevertheless, when the mammoth was finally mounted and exhibited in the Petersburg Museum of Natural History, it quickly became a world-wide sensation.

Along with the skeleton and hide, Adams had also sent a number of mammoth teeth to St Petersburg. He recognized quite correctly that 'the tusks of these archaic elephants were much more crooked and therefore much longer than those of living species'. The longest tusk Adams saw was reportedly twenty-three feet long. How the teeth were fixed in the skull was not known at the time; well into the twentieth century, in fact, they were almost always placed wrongly in reconstructions. Their proper position was not understood until cave explorers found the drawings of mammoths made by Palaeolithic men.

Adams' mammoth led to some significant conclusions. Its hairy coat, as Adams himself pointed out, was a clear proof that the animal had been equipped for life in cold regions. The same was true of another hirsute large animal that had lived with the mammoth both in Europe and in Siberia: the woolly rhinoceros. In 1773 Peter Simon Pallas first discovered the cadaver of such a rhinoceros in the frozen tundra soil of the Vilui, a tributary of the Lena. Blumenbach described the find, and realized that the animal belonged to the same species of rhinoceros that had been found in the European 'Diluvium'. Hence the animals found in the Siberian permafrost in such a remarkable state of preservation must be identical with the diluvial beasts of Europe, which were known only as fossils.

For a long time this observation sowed confusion; it could scarcely be reconciled with the prevailing theories of Creation and catastrophe. Some zoologists even speculated that mam-

Cave drawing of a mammoth, showing the proper placing of the tusks. (*From a drawing in the cave of Font-de-Gaume, Dordogne.*)

moths, woolly rhinoceroses, and other characteristic animals of the Diluvium in Europe might still be alive in remote regions of northern Asia. The problems began to clarify only after the victory of the Ice Age theory. Mammoths and woolly rhinoceroses, it was decided, must be typical creatures of periods of glaciation. Since they had adapted completely to the harsh climate and the conditions of life in the frozen tundra, they could get on quite well in Europe during times of glaciation, when the climate approached that of northern Asia. Mammoths in fact wandered as far as Alaska, where their frozen cadavers have been found in the ice.

When the rhythmic alternation of Ice Ages and interglacial periods in the last six hundred thousand years became an established fact, it was discovered that Pleistocene fauna and flora periodically shifted in the same rhythm. During the cold periods mammoths, woolly rhinoceroses, reindeer, musk ox, and other cold-loving animals lived in Europe. Whenever an interglacial period brought warmer temperatures, they migrated north and northeast. From the south the warmth-loving species would then advance, only to retreat again during an ensuing glacial period.

Early Man and the Mammoth

The discovery of the Ice Age also marked the beginning of the discovery of human prehistory. Almost at once there were bitter scientific controversies springing up over the relationships between early man and the mammoth. The first pioneers of prehistoric research, Jacques Boucher de Perches, Edouard Lartet, Henry Christy, and other discoverers of the bones and cultural traces of Old Stone Age man, repeatedly found the remains of mammoths and other Ice Age animals among the human remains. Palaeolithic strata even yielded carvings and small statues made of mammoth ivory. From these facts the *avant-garde* prehistorians drew the conclusion that palaeolithic man must have lived at the time of the Ice Age, together with mammoths and the other Ice Age animals, which had obviously been the game he hunted.

For more than half a century this view was stoutly opposed by scientific authorities. Among its most forthright critics were the great German physician and biologist Rudolf Virchow and the Danish zoologist and prehistorian Johann Japetus Steenstrup, a specialist in Arctic fauna and the prehistory of northern lands. Steenstrup reiterated again and again that Ice Age man never existed; consequently, man could never have stood face to face with an Ice Age animal such as the mammoth.

The arguments centred chiefly around one site in the vicinity of Pfedmost, a village in Moravia. The farmers here had gradually carted off a hundred-foot hill of loess that was chockful of fossils; they crushed the bones to powder which they used for fertilizer. By chance the prehistorian Heinrich Wankel heard of this. He at once appealed to the Minister of the Interior to put a ban on these depredations, and with the support of the Brünn Academy undertook large-scale excavations. Huge amounts of mammoth bones came to light, but in addition human bones of that palaeolithic culture which we now call Solutrean, and which is estimated to lie about twenty thousand years in the past.

After Wankel's death, the Czech prehistorian Karel Maska took over the project, along with a notary named Kris who was an enthusiastic scientific amateur. They discovered a large number of human tools, together with shattered and split mammoth bones. By and by similar finds were made at other places in the vicinity. Particularly interesting were the sculptures, small figures of animals and even of human beings, almost all made of mammoth ivory.

The discoverers had no doubt that they had come upon the camp sites of nomadic mammoth hunters. But Steenstrup vehemently took issue. He studied the site of Pfedmost and pointed out in a sensational account that the Tunguses and other hunting peoples of northern Asia to this day made a practice of butchering the mammoth cadavers they came upon. Why should not that have been the case in Moravia in the past? According to Steenstrup, the hunting tribes of Pfedmost had not hunted mammoths, but only dug frozen cadavers of mammoths out of the ice – corpses that had been kept in the cold storage locker of the frozen ground for ages. They had used the meat, cracked open the bones, and made ivory ornaments out of the tusks, even as the dwellers of the Siberian tundra still did.

But Steenstrup had bad luck. Shortly after his study tour in Pfedmost in 1895, the notary Kris found a small figure of a mammoth that a Stone Age hunter had carved out of a mammoth tusk. It was a perfect likeness; even the large fatty hump could be seen on it, a feature of the mammoth that zoologists had not suspected. The existence of the hump was later confirmed by the discovery of the cave paintings after 1900.

Thus Steenstrup's theory collapsed completely. Subsequently, when innumerable drawings of mammoths were found in the caves of Combarelles, Font-de-Gaume, La Mouthe, Madeleine, Pech-Merle, Rouffignac, and many others, there could no longer be any doubt that man and mammoth had been contemporaries during the Würm (Wisconsin) Ice Age. Palaeolithic man had shown the mammoths individually and in herds, had made like-nesses of bulls, cows, and calves, sometimes attacking, some-times captured in artful pitfalls, sometimes stuck with arrows: a bit of prehistoric hunting magic. The ancient cave painters had been excellent observers of nature; with striking realism they had rendered the shaggy coats of the huge beasts, the upthrust head, the huge hump on the back, and the characteristic position of the curving tusks.

Science must be grateful to the artists and medicine men of the Late Ice Age. For the paintings, scratch drawings, sculptures, and carvings of that earliest epoch of man's art also depicted the many other animals which lived during the glacial periods and the warm intervals: the woolly rhinoceroses, steppe bisons, aurochs, wild horses, stags, cave lions, and bears. Had we not had this prehistoric art, our notions of the Ice Age fauna would be extremely incomplete.

On one point, however, the sceptical Steenstrup was after all proved right. Ice Age man was a mammoth hunter; that is beyond doubt. But had he also killed the herds of mammoths of Pfedmost? Probably not. The remains of from five to six hun-dred mammoths have been excavated from that one site. It is hard to conceive that man with his then inadequate weapons could have wiped out such a huge collection of gigantic and by no means defenceless animals. Of course the palaeolithic hunters could gradually have killed so many mammoths here and there in the Moravian plain over a long period of time. But if so, they

Lion or tiger? In this prehistoric drawing the cave lion looks like
a tiger. Since the two can scarcely be distinguished by their
skeletons, we do not know for certain whether there were two
species of large predatory cats in Europe during the Ice Age, or
only one species which looked like a lion, a tiger, or a 'lion-tiger'.
(*From the cave of Combarelles, Dordogne.*)

would scarcely have been able to drag the colossal cadavers any
distance and assemble them all at this one spot.

The only remaining possibility, it would seem, is to assume
that the mammoths of Pfedmost were driven together into one
gigantic herd by some natural disaster, perhaps an unusually
protracted blizzard, and then died on the spot. Wolves and cave
hyenas would soon have discovered the mammoth carcasses
and thus led the hunting nomads to the spot. The men must have
dug the bodies out of the snow and made use of the meat until
it was no longer fresh.

Ice Age Graveyards

Sizeable accumulations of mammoth bones or mammoth teeth
have been discovered not only at Pfedmost, but also in the loess
regions of Austria, in Württemberg, near Lake Constance, in
northern Asia, on the New Siberian Islands in the Arctic Ocean,
and even at the bottom of the North Sea. In the years between

1820 and 1833 oystermen in the vicinity of the Dogger Bank dredged up more than two thousand mammoth molars out of the ooze. In Europe there have so far been found, *in toto*, the bones or teeth of approximately twenty thousand mammoths, in Siberia of nearly fifty thousand individuals. To these must be added several thousand mammoths from North America.

The extraordinary frequency of mammoth bones in certain regions, such as Moravia, in the neighbourhood of the Dogger Bank, and in the barren New Siberian Islands, led Wolfgang Soergel in 1912 to postulate that these places had been huge graveyards to which sick or dying mammoths constantly resorted. Soergel referred to the alleged 'elephant graveyards' of Africa, which big game hunters had so often reported.

But the elephant graveyards have by now been consigned to the realm of fable. Why mammoth bones and mammoth teeth have so often been found in amazing concentrations remains a mystery. Even if we assume that these bones accumulated over great stretches of time, the question remains why the mammoths were in these places in such great numbers, and not more evenly distributed over the earth.

In Siberia most of the teeth and almost all of the deep-frozen cadavers have been found in the coldest, most barren and inhospitable areas of the Northeast. Today we know that the mammoths lived in Siberia and Alaska far longer than they did in Europe. Whereas they died out in Europe towards the end of the Würm Ice Age, they formed separate species in Siberia and Alaska, and continued to wander up and down the tundra deep into the postglacial era.

In the Siberian tundra they found the same conditions as they had experienced in Ice Age Europe. They fed chiefly on grass and tundra plants, on reindeer moss and the sprouts and young growth of conifers. Consequently, zoologists and palaeontologists repeatedly asked themselves why the mammoth (and the woolly rhinoceros as well) did not go on living to the present. The reindeer and the musk ox, two other typical Ice Age animals, drifted northward and north-eastward in the postglacial era and inhabit those cold regions to this day.

It may be that large animals like the mammoth and the woolly rhinoceros suffered especially from the unusually heavy precipitation of the postglacial period, and for that very reason sought out the most barren regions of northeast Asia – which also have the lowest precipitation – as their last refuge. After the withdrawal of the Würm ice sheet, Europe was subject to violent rainfall, which must have been uncomfortable to animals accustomed to the dry, almost germ-free polar air. In the mountain regions avalanches formed. The Siberian tundra was swept by snowfalls and blizzards of unusual proportions. Snow covered the ice of rivers and lakes to a thickness of many yards; and if the ice beneath was thin, a large animal could easily break through because it had had no previous experience with such dangers. Pits and crevasses filled with snow similarly became veritable animal traps.

The fate of such animals has been dramatically demonstrated

The warmth-loving jungle elephant was also depicted by men of the old Stone Age. Abbé Breuil would have it that the heart-shaped mark on this Spanish cave drawing was meant to show where the primitive elephant hunters were to aim their spears. According to other scientists, the 'heart' on this famous painting is merely an accidental red spot. (*From the Pindal cave, Asturias.*)

by the finds of mammoths and woolly rhinoceroses in the Siberian ice. In August 1900 a Lamut tribesman named Tarabykin discovered on the bank of the Beresovska in north-eastern Siberia an extremely well-preserved mammoth cadaver, which was obtained a year later by an expedition sent out by the St Petersburg Academy under the direction of two zoologists, Otto Herz and E. W. Pfizenmayer. Herz and Pfizenmayer identified the plants on which the animal had fed from the stomach contents and unchewed remains of vegetable matter caught in the teeth. They also reasoned out that the mammoth had been caught suddenly by death in the early fall, and could tell just how it had perished.

Freshly fallen snow had completely filled a deep crevice in the ground, turning it into a pitfall. The mammoth had wandered into this trap just as the elephants of the African forests stumble into camouflaged pits dug by natives. In falling, the mammoth had broken its pelvic bone and right foreleg.

Such accidents apparently were by no means unusual. A mammoth discovered by another Lamut named Dyakov in 1907, by the Sanga-Jürach River, and a woolly rhinoceros dug out of the

Ice Age man must have known the huge beasts well, as this representation of a fight between two mammoths shows. (*From the cave of Les Combarelles, Dordogne.*)

ice by Gorochov in 1877 on the steep bank of the Chalbui River, had met death in a similar manner. In America the cadavers of mastodons – members of another species of proboscidian

– have repeatedly been found in much the same positions and circumstances as the dead mammoths. The mastodons had also been killed by sliding down steep slopes or falling into crevasses.

Swamps, asphalt lakes, and deposits of ozocerite (mineral wax) frequently turned into animal traps. The Polish palaeontologist Niezabilowski, examining an ozocerite deposit near Starunia in Galicia, found the mummified cadavers, completely soaked with the mineral paraffin, of two woolly rhino calves and one mammoth calf. But such individual accidents, although they might happen more frequently under some environmental conditions, would not suffice normally to cause the extinction of a species. Such accidents become calamitous only when the species has already been weakened or in some way made unfit for survival.

Was that the case with the mammoths and other Ice Age animals that have disappeared from the earth? Did disease and degeneration cause or accelerate the extinction of these creatures? That problem has become the province of one of the most interesting branches of palaeontological research – the study of 'fossil diseases'.

The Mystery of Dying Species

With the keenness and devotion of detectives, palaeontologists have sought to determine why some species of the Ice Age died out, while others did not. Among the extinct species, in addition to the mammoths, mastodons, and rhinoceroses, are certain herbivores such as the giant deer and the broad-browed elk, as well as several carnivores: the cave bear, cave lion, and cave hyena. They constitute a minority, for the greater part of the Ice Age fauna are still living among us today, unchanged or scarcely changed since the days of the great cold. In the far north, along with reindeer and musk oxen, the wolves, lynxes, wolverines, Arctic foxes, snow hares, and lemmings have survived the passage of ages. In Central Asia the wild horses, the dziggetai or half-ass and saiga antelopes lived on. The ibexes, chamois, and marmots withdrew into the high mountains. Present-day bison are not identical with the Ice Age steppe bison, but closely akin to them. The aurochs and wild horses were the ancestors of our present-day domestic cattle and horses.

Why did several species die out, while others could hold on? Many possible explanations have been advanced: the dire results of climatic changes, epidemics, even extermination by Stone Age man. Finally there was the endeavour to interpret the phenomena of prehistoric life and death by comparisons with the present. The Austrian scientist Othenio Abel took this approach when, in 1912, he founded the new discipline of palaeobiology.

The palaeobiologists set out to determine what prehistoric animals looked like, how they lived, what was the character of their environments, and for what reasons any given species died out. From comparative analysis of the bones of fossil and living forms it proved possible to deduce the manner of life of extinct species. In addition, there is a surprising amount of evidence bearing on the habits of extinct creatures: trails, traces of crawling and scratching, fossil remains of food and excrement, burrows, wounds, injuries, indications of disease, skeletons in the distorted positions of death agonies. A further field of research necessarily developed: palaeopathology, the study of fossil diseases. Many fossils revealed that the animals during their lifetimes had suffered from fractures, caries, gout, cancer, rickets, inflammations of the jaw, and other diseases. Avian osteopetrosis, a disease of domestic fowl, has even been diagnosed in dinosaurs. If morbid changes and severe deformations occur frequently, that is, if a great many individuals obviously succumbed to such illnesses, that fact might throw some light on the reasons for the vanishing of a species.

Degenerate Mammoths

Fossils also furnished information on the processes of degeneration in a number of prehistoric species. The record of the mammoth proved especially illuminating. We ordinarily think of the mammoth as a gigantic beast, and that image holds true for the ancient mammoths of the first periods of glaciation; the animals were truly of 'mammoth' proportions. During the Mindel (Kansan) Ice Age the bulls were nearly seventeen feet high at the shoulder, and in the Riss (Illinoian) Ice Age still fifteen feet. But the later forms, from which in fact we derive our picture of the animal, attained nothing like such size.

'As early as the beginning of the Würm (Wisconsin) Ice Age,' Abel points out, 'the mammoths showed certain signs of degeneracy.' The bulls by then were only ten feet high, and the cows a mere eight and a half feet – smaller than present-day elephants. In some regions they dwindled still further. Abel calls the mammoths in the loess of Krems on the Danube 'dwarfs' – a rather paradoxical-sounding term.

Diminution in size is sometimes a sign of degeneration and lessened fitness for survival. We may imagine that mammoths in their degenerated, shrunken condition were no longer able to cope with the changes of climate and environment in the post-glacial age. Hence it is quite possible that they gradually died out from accidents or epidemics of disease which in the past they had easily survived.

So far, no obvious signs of disease have been detected in mammoth bones. Another giant animal of the Pleistocene has, however, given a clear accounting to palaeopathologists. This was the huge stag whose shovel-like antlers spread to widths up to ten feet. This giant deer assumed particularly stately forms in Ireland during the Ice Age. Another variety lived in central and eastern Europe. In the Black Sea region it lived on for some time into the postglacial age and possibly – if certain drawings of stags in Scythian graves have been correctly interpreted – into the Bronze Age.

This giant stag had each year to grow a pair of antlers weighing up to eighty pounds; every winter it cast its antlers just as the red deer, the elk, and related species do today. 'Such physiological events are conceivable and possible only if there is a peculiar production of hormones,' Abel emphasizes. And it is well known that intense hormone activity quite often leads to conditions which can become chronic or even fatal.

One of these conditions is a severe thickening of the lower jaw. It is called pachyostosis, and is occasionally encountered today in stags that have especially heavy antlers. Among the giant stags, pachyostosis became chronic; judging from the jawbones that have been found, each and every animal displayed it. Thus these colossal deer presumably died out from overdevelopment of their most distinctive feature.

Bear Hunting and Bear Cult

Even more illuminating was the pathology of the gigantic cave bear, which Abel studied with extreme care. In older books on palaeontology the cave bear was described as a grim predator, a serious enemy of Ice Age man. In reality the huge beast lived mostly on vegetation and probably ate meat only now and then. The nature of its diet may be read from the way in which its teeth were worn down and also from the contents of the stomachs of several cadavers preserved in alpine caves under yard-thick layers of bat droppings. In addition, bones of other Ice Age animals frequently show traces of bites from wolves and cave hyenas, but never from cave bears. Furthermore, the severe deformations of the jaw in a surprisingly large number of individuals are further evidence for the herbivorous nature of the cave bear. The cause of this disease, known as actinomycosis or lumpy jaw, is a ray-fungus which attacks exclusively herbivores.

In France, England, and Central Europe, but particularly in the caves of the alpine regions, innumerable remains of cave bears have been discovered. Thus this animal is especially familiar to scientists; its life and its extinction can be reconstructed most convincingly. We have even learned what drove the shaggy colossus into underground recesses. The French cave-explorer Norbert Casteret had this to say after studying the Pyrenees cave of Pène-Blanque:

Clear traces have been preserved in the clay, permitting us to reconstruct one of the most remarkable scenes in the life of the cave bear. The bears used the natural conditions of the caves for sliding parties, the cave floor forming a great chute which ended in the muddy water. Many of the tracks are so clear that we even see the impress of hairs from the pelt upon the soft clay. . . . In one corner a bear overcame ennui by dancing on the spot for hours, as the many overlapping imprints of its paws on the clay floor testify.

Some of the caves show that the human contemporaries of the cave bears, the Neanderthal men, caught the big animals in nooses and killed them by striking them on the head with sharp-edged weapons. Innumerable paw and claw marks on the cave walls memorialize the bears' struggles. Sometimes severely wounded animals were able to escape from the hunters. Abel describes a cave bear skull from the dragon cave near Mixnitz in Syria which bore the mark of a heavy blow. Suppuration channels in the frontal bone indicate that the wound never completely healed, but that the bear nevertheless survived, probably for many years.

The caves show traces not only of prehistoric bear hunts, but also provide amazing proof of the part the great bears played in the religious life of Neanderthal man. Above Vättis in the Tamina Valley, eight thousand feet above sea level, there is a 'dragon cave' in which the Swiss prehistorians Bächler and Nigg found large numbers of cave bear bones from 1917 to 1921. The skulls and bones were laid in niches or stone chests, were arranged in deliberate order and carefully covered with stone slabs. Bächler and Nigg discovered that the Neanderthal men had decapitated the bears and then buried the heads whole. Sometimes thighbones were thrust through the cavities in the skulls.

Similar finds have been made in other caves. The Franconian Petershöhle, near Velden, the Swiss Wildmannslisloch, the Reyersdorf cave in Silesia, were all cult sites of Neanderthal men. These primitives established collections of bear skulls in the caves; probably they also spitted the heads of bears on poles and performed solemn dances around them. Such bear ceremonies and bear sacrifices occur to this day among the Ainu, the Gilyaks, and other peoples of northeast Asia.

These discoveries have led to the supposition that man may have been responsible for the extermination of the cave bear. But there are no proofs. The cave bear survived into the last glacial period. It died out during the palaeolithic cultural epoch known as the Solutrean. The number of human beings in those times was too small and their weapons were too rudimentary for them to have represented a threat to the survival of any animal species. It was not until much later, when man had formed all sorts of prideful ideas about his special intellectual and moral position in the universe, that he began his campaign of annihilation against the creatures of nature.

The Animal That Domesticated Itself

The cave bear fell victim to another, highly significant phenomenon of the natural world. Indeed, its fate is probably typical of that of other extinct species. As Othenio Abel has put it: 'Almost always a species or group of species experiences a period of florescence which is followed by a more or less rapid decline. We need only think of the history of the mammoth, the mosasaur, and many other groups.'

The more a species dominates its environment during its period of florescence, the more vulnerable it is to crucial disturbances of its normal vital functions. Mutations are constantly appearing in all living organisms, changes in the genetic stock which need not necessarily be advantageous. In general, nature permits only the useful mutants to survive, and extinguishes the useless or baneful genetic changes in the course of the struggle for life. But where natural selection fails, useless or negative mutations proliferate; innumerable variants form, much as in our domestic animals, which have long been deprived of natural selection. Since, however, such typical results of domestication are largely useless or disadvantageous in natural conditions, the species becomes as a result more and more unfit for survival.

The cave bear experienced its florescence during the mild Riss-Würm (Sangamon) Interglacial Epoch. Unlike the mammoth and the woolly rhinoceros, it could exist in both warm and cold periods; it did not migrate when the climate changed, but tried to adapt to the new conditions. In the Riss-Würm interglacial it developed into a gigantic animal ten to twelve feet long and over five feet high at the shoulder. It was thus the largest bear that ever lived.

It owed its size to optimum living conditions. The summers were long; there was plenty of grass and other provender. It did not have to hibernate for many months. Above all, however, it had virtually no enemies; it could easily fend off the big cats, such as the cave lions, and during the interglacial period man hunted it either very rarely or not at all. Yet this period of flourishing prosperity seems to have ushered in its ruin.

As the Viennese zoologist Otto Antonius was the first to perceive, the cave bear 'domesticated itself like man' during this period. The huge beast was no longer exposed to the struggle for existence. Natural selection scarcely operated among the cave bears. Thus they were prone to virtually the same types of mutations that were experienced by the domesticated dog in the course of its history.

The struggle for existence is often taken by those unversed in natural history as a sign that nature is cruel. In reality the close selection it imposes is a blessing to living organisms, it guarantees the continuance and higher evolution of the species. In the absence of natural selection, dwarfed, shaggy-haired, stubby-legged, feeble or degenerate individuals not only live much longer but cross constantly with normal and healthy members of the species. Thus, as Abel describes it, 'in the course of generations the whole species increasingly deteriorates; and this process continues until the enfeebled descendants are no longer able to cope with some new disturbance in the environment'. In other words, the species dies out of degeneration.

Such degeneration apparently overtook the cave bears. The period of florescence was followed by the Würm glaciation with its unusually long and severe winters. Since the cave bears could not migrate, they were compelled literally to sleep away the greater part of their lives. From examination of innumerable cave bear bones Abel discovered the consequences for the already degenerate animals:

The long winter meant virtually a prison existence for two-thirds of the animal's life. It is not surprising that this prolonged internment had extremely harmful effects upon the health of the bears. Among brown bears and other large predators which are kept for many years in the confinement of cages, peculiar diseases of the spinal column develop. The vertebrae are attacked by inflammations; fusing and hypertrophy of the bones appear. Almost invariably two adjacent vertebrae of the posterior thoracic and anterior lumbar region are affected by this 'prison disease'. The same vertebral afflictions are well known from many cave bear sites. The cave bears, too, suffered from having to spend some two-thirds of their lives in caves like prisoners.

In addition to vertebral diseases and actinomycosis, Abel and other scientists found that the cave bears had fallen prey to many other diseases: rickets, gout, arthritis, tuberculosis of the joints, inflammation and atrophy of the bones. The teeth were especially affected; they had such little use during the long winters that, since they grew continually, they actually blocked the mouths of some individuals. A surprisingly large number of cubs, newborn bears, and embryos have been found in the caves. Kurt Ehrenberg and other investigators deduce from this fact that epidemics must have raged through the cave bear population, particularly affecting the young animals. In addition, frequent occurrence of miscarriages and premature births points to a weakening and degeneration of the female bears, who were no longer able to carry their offspring to term.

Unquestionably, the Neanderthal men and the men of the Solutrean culture wrought havoc among these sick, defective, and crippled cave dwellers. The bears were especially vulnerable during their hibernation, and it appears that men stunned them by smoke and blinded them with torches when they were trapped in the caves, so that the animals could more easily be slain in narrow passageways. Even so, the human hunters would not alone have succeeded in utterly annihilating these sickly colossi. Ultimately, we may say, the cave bear died as a result of its own self-domestication, of the degeneracy, the unfitness for survival, which seems to take place inevitably whenever a species has been for some time – to put it in the words of Lorenz – 'the lord of the earth in much the way that man is'. Nature herself destroyed the cave bear because she had previously been so cruel as to spare it the 'cruel' struggle for life.

Do We Live in an Interglacial Age?

For the past twelve to fifteen thousand years the last interval of cold in Central Europe and America has gradually been abating. But does that mean that the Ice Age is ended once and for all? Or are we again heading for a time of glaciation which may overtake us in a few thousand or tens of thousands of years?

If we examine Milutin Milankovich's radiation curves, we may well be gripped by doubts about the future. For these curves show us that every forty to one hundred thousand years cosmic conditions are favourable for a cold age. Does this mean that we live in an interglacial age, a warm interval between two advances of the great ice sheets? There is reason to think so. According to the American physicist George Gamow, there is a good deal of evidence supporting such a view.

As Gamow points out in his *Biography of the Earth*, the astronomical preconditions for a glacial age recur throughout the entire geological history of our earth at intervals of less than one hundred thousand years. But glaciation can take place only in geological epochs in which high mountains cover much of the surface of the continents. The existence of mountains is the indispensable prerequisite for the origination of glaciers and hence for the onset of an ice age.

The process is much the same each time. First thick layers of ice form on the heights. If astronomical factors cause a general cooling of the climate by only a few degrees, the layers of ice constantly increase; under their own weight the glaciers slide down from the mountains and ultimately cover extensive areas of the lowland. Today some fifteen million square kilometres of the globe are covered with ice. During the glacial periods of the Pleistocene, however, the average glaciation was fifty-five million square kilometres, almost four times as much. This ice came solely from the glaciers in the high mountains.

In times of diminished orogeny (mountain building), therefore, there are no glaciers and consequently no icing of the plains. If astronomical factors produce a lowering of the average temperatures during such epochs, there is no dramatic consequence such as glaciation. At most some zones which had previously been tropical turn subtropical and other zones, hitherto subtropical, have temperate climates.

If we wish to discover whether we are approaching a new Ice Age, we must first determine to what extent a lowering of the temperature today would result in more active glacier formation.

Mountain Building and the Ice Age

Gamow has made a number of observations on our present situation which are not exactly consoling. He points out that we are living more or less at the midpoint of one of our planet's 'creative' periods, during which mountains are forming. A number of high mountain chains already exist, and the formation of more seems to be impending. We may therefore expect, Gamow argues, that the ice will soon be returning, and that periodic advances and retreats of ice will continue as long as there are still mountains in northern latitudes.

This prognosis loses some of its terrors when we realize that geologists take a view of time rather different from that of ordinary mortals. By 'impending' and 'soon' Gamow means not the immediate future, but several tens of millennia ahead. We have a little breathing space before the next return of the ice.

The connexion between orogeny and ice ages can in fact be demonstrated throughout our planet's history. In the history of life on earth, this phenomenon has played a significant part. While the orogenies took place at various times in different places over vast periods, a general chronology may be described.

The first high mountains of which we have definite knowledge were formed in the Algonkian, about six hundred million years ago. They vanished long ago. But when they still towered into the sky, during the Precambrian and Cambrian, the first detectable Ice Age on our planet descended upon Australia and the Far East. During this period almost all the phyla of invertebrate animals flourished in the seas. Can it be that the Ice Age had some bearing on the florescence of invertebrate life?

A hundred million years later came the Caledonian orogeny. So far as we can tell, it led to local glaciations in Scandinavia and South Africa during the Silurian period. But it was in the Silurian and the following Devonian that plants and animals gradually conquered the dry land. Again, the question arises whether this most significant event in the history of life may not in some way be connected with the upthrust of the mountains.

The Caledonian orogeny entered a phase of tremendous mountain building, what is known in geology as the Appalachian Revolution, which took place largely during the Carboniferous period. It was followed by the Permian Ice Age, probably the most extensive period of glaciation in the history of the earth. Almost the entire southern hemisphere was covered by one vast ice sheet. Towards the end of the Permian and the beginning of the following Triassic period the first primitive mammals emerged from mammal-like reptiles. Perhaps the Permian Ice Age had something to do with the appearance of these first warm-blooded animals.

In the Mesozoic, between 200 and 60 million years ago, the Caledonian and Appalachian mountains were gradually worn down. Many middle-sized mountains of the present days are modest remainders of the Appalachian Revolution: in North America, the Appalachians; in Europe the Black Forest, Thuringian Forest, and Ore Mountains. But while these mountains were being reduced, the first phases of Cascadian-Alpine mountain building began in Triassic, Jurassic, and Cretaceous times. At first no huge mountain ranges arose. Nevertheless, towards the end of the Cretaceous there seem to have been minor periods of local cold. That was the age in which the giant saurians died out and more and more species of birds and mammals appeared. It seems likely that there were again some links between geological and biological events.

The Alpine upthrust reached its climax in the Tertiary, that is during the last sixty million years. This was when the Alps, the

Pyrenees, the Carpathians, the Andes, and the Rocky Mountains were formed, as well as the Caucasus, the Altai, and the Himalayas. What resulted was the 'classical' Ice Age of the Pleistocene, with its alternations of glacial and interglacial periods, its mammoths, woolly rhinoceroses, cave bears, and other fauna innured to cold. Towards the end of the Tertiary the first precursors of man appeared; in the Pleistocene they gradually developed into skilled toolmakers, hunters, artists, and weaponsmiths; into thinking, rational, creative, brain-directed animals. Anthropologists almost unanimously hold the view that the harsh conditions of life during the last Ice Age contributed considerably to the rapid rise of *Homo Sapiens*.

The Alpine orogeny is still in full swing at the present time. It is quite possible, Gamow observes, that all the many earthquakes and volcanic eruptions which can be traced throughout the entire known history of mankind were only preparations for the next great upthrust of new mountain ranges. The fact that we live in such a phase need not, however, lead us to conclude that any day new mountains may spring up on the surface of our earth like mushrooms after a summer rain.

The fact remains that existing mountains and those in process of formation are sufficient to being about a new advance of the glaciers the moment the eccentricity of the earth's orbit establishes the right cosmic conditions.

Presumably, then, we live in an interglacial period corresponding in general to the interglacials of the Pleistocene. The question of great importance for us is, therefore: is this warm period just beginning – or is it approaching its end?

It is gratifying to learn that most geologists think the present interglacial period has just begun. Probably the average temperature in the coming millennia will continue to rise and the climate become even milder and more uniform than it is today. Perhaps a period considerably longer than the whole of human history so far will pass before the next period of cold descends.

Since palaeolithic men, in spite of their pitifully small equipment, managed so well to cope with the unusually cold climate and heavy glaciation of the Würm Ice Age, technologically advanced civilized man will certainly withstand the test of any new advance of the glaciers – assuming that at this time, a few tens of thousands of years hence, any of the human race still occupies this planet.

But whether it will or not probably does not depend on geological factors, but primarily upon ourselves.

Book Four

The Family Tree of Life

At last gleams of light have come, and I am almost convinced (quite contrary to the opinion I started with) that species are not (it is like confessing a murder) immutable.

CHARLES DARWIN, 1844

The Mystery of Mysteries

On 27 December 1831, a twenty-two-year-old naturalist began his scientific career by taking a trip around the world. When the young man returned to his home in England five years later, he brought back with him such abundance of material in the form of collections, observations, and experiences that he was able to draw upon these riches for the rest of his life. Not only that: this round-the-world tour supplied the young scientist with proofs of a new theory – a theory he released to the world at large a quarter of a century later. The result was an instantaneous thunderclap and a scientific, intellectual, and social chain reaction of revolutionary proportions.

Even in the eyes of his opponents, Charles Darwin is regarded as one of the great figures in our cultural history. Along with Marx, Freud, and Einstein he belongs among the fathers of the modern scientific age. His theory of evolution had an enormous impact upon the most varied regions of human life and culture. It radically recast the picture of man and the universe that had been generally accepted before its promulgation. In almost all disciplines it shook or overthrew existing tenets, dogmas, and doctrines.

Darwin was not the first evolutionist in his family. His grandfather Erasmus Darwin, physician, naturalist, and poet of the Baroque Age, had expressed evolutionistic views in a number of books and didactic poems. Some of his thought is reminiscent of Lamarck, some of Goethe. Such a cult was made of Erasmus Darwin's ideas within the family that it may well be asked whether Charles was not preordained to take up the cause of evolution.

Strangely enough, that was not the case. Darwin rejected Lamarck's highly speculative tenets about the changes environment could wreak on species as vigorously as he rejected the element of natural philosophy in Goethe's ideas. This alone put him on the outs with his grandfather. In fact he himself commented that it was remarkable to what extent his grandfather 'anticipated the views and erroneous grounds of opinions of Lamarck'. Charles Darwin was not interested in philosophical speculations, but in unequivocal, demonstrable facts. For that reason he was successful, unlike all his predecessors, and for that reason the idea of evolution is now inseparably connected with his name.

The *Beagle*, on which Darwin was to travel around the world, was a research and survey vessel of the British government. Its captain, Robert Fitzroy, was still quite a young man, but already a highly respected cartographer and 'meteorologist who was to carry out a wide variety of scientific missions in South America and the Pacific and Indian oceans. Among other things, he was to survey coasts, measure currents, work up land and naval maps, study climatic conditions, and assemble extensive geological, botanical, and zoological collections.

He needed an assistant who would undertake the requisite excursions and keep a record of the scientific observations. For an ambitious and talented young scientist here was a chance to qualify as an independent researcher. Fitzroy appealed to John Steven Henslow, the botanist and geologist, to recommend a capable man for the *Beagle*.

As chance would have it, Professor Henslow's favourite pupil was young Darwin. At this time Darwin had by no means settled on his future vocation. He had had to give up the study of medicine, prescribed by family tradition, because he could not endure the sight of blood. In Cambridge he studied theology and took his B.A.; but he was disinclined to enter the Church. So far he had shown no special talent for scientific work. But Henslow had seen, on a number of excursions into the Cambridge marshes, that young Charles might make a good biologist if given the chance to develop his slumbering faculties. And so he decided to propose Darwin for the post on the *Beagle*.

In spite of Henslow's recommendation, the opportunity very nearly came to nothing. That was due, absurdly enough, to Darwin's nose. Robert Fitzroy was a follower of Lavater and apt to judge people by their physiognomic traits. The young Cambridge student's nose betrayed, it seemed to him, a soft and indecisive character. What was more, they were on opposite sides politically. Darwin was a dyed-in-the-wool Whig, Fitzroy an equally strong Tory. After prolonged hesitation the conservative captain was persuaded to take the liberal naturalist with him, in spite of his compromising nose and his incompatible politics.

As it turned out, the two men got along famously. Fitzroy quickly saw that Darwin carried out his tasks of collection and observation with the perseverance and reliability of a fully mature scholar; the captain could leave the whole field of biology to the young man without any second thoughts. Darwin, on the other hand, soon demonstrated to Fitzroy that he was anything but soft and irresolute. In South America Darwin witnessed with horror the effects of the slave system; and when Fitzroy defended slavery, Darwin turned on him so passionately that the other members of the expedition feared a final breach between the two. Instead, they argued their way to friendship and learned to respect one another. In fact shortly after the dispute Darwin wrote to Henslow thanking his good fortune that Fitzroy had not made him an apostate to the principles of the Whigs. He did not wish to be Tory, if only because of the indifference to this scandal for Christian nations – slavery.

There were other occasions during this voyage when Darwin took the stand of a deeply religious Christian. It must not have entered his head that he would some day be denounced as the father of materialism and atheism. At the beginning of the voyage, as we have seen, he was still a loyal follower of Cuvier and believed firmly in the immutability of species. His faith in the prevalent scientific views was first shaken during the voyage, when he read a book on geology which – it may be added – did not so much as mention the possibility of organic evolution. The book was Lyell's *Principles of Geology*.

The first volume of the work had been issued just before Darwin set sail, and Henslow sent it to his pupil before embarkation. He mailed the second volume as soon as feasible. Henslow was by no means an adherent of Lyell. On the contrary, he warned Darwin against these theories. The work, he wrote, was certainly highly interesting, but Darwin must be careful not to swallow all of Mr Lyell's assertions.

Henslow could not suspect that the *Principles* were to strike Darwin with the force of revelation in the course of his long voyage. They provided the decisive impetus for the young scientist's own observations and reflections. They shook Darwin's confidence in the tenets of Cuvier. Henceforth Darwin looked at the geological conditions in South America with Lyell's eyes; he saw the stratification, the slow rises and falls of land, the growth of mountains, the effects of the Ice Age. He put together a wealth of material that corroborated Lyell's prime principle: that the face of our planet in former ages had been modelled by the same forces that were still acting in the present.

But if Lyell believed that the aspect of the earth had evolved in the course of tremendous aeons, it necessarily followed that its flora and fauna had also evolved. Thus early Darwin began toying with the idea of evolution, and struggled against the temptation to elaborate it. For the time being he resisted; his distaste for philosophical speculation was too strong. Before he dared to attack Cuvier's theory, he had to command sufficient proofs; he had to amass facts.

The Enchanted Islands

In his travel journal and his later writings, Darwin repeatedly stated that he had been led to his theory of evolution chiefly by his observations in the Galápagos Islands. When in October 1835 he began studying the peculiar fauna of this remote archipelago, he finally realized that the present forms of plants and animals could only be accounted for by evolution: 'Hence, both in space and time, we seem to be brought somewhat near to that great fact – the mystery of mysteries – the first appearance of new beings on this earth.'

The Spanish discoverers had called the Galápagos *Las Encantadas*. They are of volcanic origin, situated approximately five hundred miles west of Ecuador, in the Pacific Ocean. To the naturalist they remain still enchanted and enchanting islands. Geologists believe that the Galápagos emerged about sixty million years ago, towards the end of the Cretaceous. After they rose out of the sea, they were for a time hot, unfruitful, and totally without life.

But these conditions changed. On the 'Enchanted Islands' Darwin saw types of fauna that did not exist elsewhere: giant tortoises, after which the islands were named, marine lizards, blunt-snouted iguanas, black finches, flightless cormorants, and a wide variety of other species and subspecies. How had all these animals reached the Galápagos? They could only have come from South America, by air or on drifting tree trunks – like the plant life which had spread over the islands. But in that case, by accepted doctrine they ought to have been identical with the corresponding species on the mainland of South America.

Darwin at once saw that this was not so. The species on the Galápagos resembled those of the mainland, but they were by no means identical. They looked like curiously specialized descendants of mainland immigrants which had adapted to the particular conditions of life on these remote islands. The giant tortoises, in fact, had formed a special race on each of the larger islands. All this, it seemed to Darwin, could only be explained by assuming that species gradually changed. The question began to obsess him.

The most astonishing creatures on the Galápagos to Darwin were not the enormous tortoises or the lizards, but the finches – the Darwin finches as they are called today. Their nearest relatives live on the mainland of South America. Since with the exception of the mockingbird no other songbirds have reached the Galápagos, the Darwin finches occupied a wide range of territories and favoured extremely variegated foods. In the Galápagos they played the part of grosbeaks and flycatchers; they behaved like titmice, nuthatches, and even woodpeckers.

Some ate pits and seeds and had developed fairly strong conic beaks; others with far frailer beaks had become leaf eaters; still others were insect eaters or even specialized cactus eaters. The type which imitated the woodpecker's habits pecked holes in trees to procure insects; but since it lacked the woodpecker's long tongue, it picked up a cactus needle in its beak, introduced the needle into the holes and thus spitted its prey: one of the rare cases in which a bird makes use of a tool.

All these Galápagos finches, Darwin reasoned, may have descended from a single pair that had been blown to the archipelago. The descendants of this pair spread over all the islands, conquered a wide variety of environments, and changed to meet the new conditions. Gradually the varieties developed into independent species which could no longer interbreed. Some types are in the middle of this process of transformation to this day. They can still breed with closely related types; they are, as a modern evolutionary biologist would express it, 'borderline cases between genus and incipient species', and thus living proofs that the development of species has far from ceased, that it is still going forward continually in our present age.

Darwin had ceased to doubt that evolution was a reality. He was approaching nearer to the 'mystery of mysteries', not only through observations on the geographical distribution of animals, but also through palaeontological excavations. For even before visiting the Galápagos he had developed into a highly knowledgeable palaeontologist. 'Had been greatly struck . . . on the character of South American fossils and species on Galápagos Archipelago,' he subsequently noted. 'These facts (especially latter) origin of all my views.'

'This Wonderful Relationship'

With Fitzroy's permission, Darwin left the ship as often as was feasible. While the *Beagle* sailed along the coasts to take soundings, Darwin rode over the pampas, bought fossils from the colonists or dug them himself out of the Tertiary clay and marl strata of Patagonia. During the long, slow days at sea he studied what he had collected.

The evidence pointed to strange and to some extent gigantic mammals having lived in Patagonia during the Tertiary period. Some of them showed not the slightest kinship to any present group of organisms; thus they seemed to corroborate the theory of geologic revolutions and successive re-creations of life. Others, however, were obviously gigantic relatives of present forms; their existence bore out the idea of evolution.

Proceeding with the greatest care, Darwin began drawing up a general inventory of the fossil fauna of Patagonia. There was the toxodon, which Darwin called one of the strangest animals ever discovered:

In size it equalled an elephant or megatherium, but the structure of its teeth . . . proves indisputably that it was intimately related to the Gnawers, the order which, at the present day, includes most of the smallest quadrupeds: in many details it is allied to the Pachydermata: judging from the position of its eyes, ears, and nostrils, it was probably aquatic, like the Dugong and Manatee, to which it is also allied. How wonderfully are the different Orders, at the present time so well separated, blended together in different points of the structure of the Toxodon!

Another bizarre beast was the *Macrauchenia*, 'a remarkable quadruped as large as a camel', whose skeleton Darwin found in the red mud near Port San Julian. This animal, as Darwin pointed out, belonged to the same thick-skinned group as the rhinoceros, the tapir, and the fossil *Palaeotherium*; but in the structure of its bones and neck it showed a distinct kinship with the camel, and even more with the guanaco and the llama.

But these resemblances to forms of present-day animals were only deceptive. In reality the *Toxodon* and *Macrauchenia* had no living relatives or descendants. They belonged among those specifically South American ungulates which the American palaeontologist William Berryman Scott has called 'mysterious groups', and which are surely among the strangest and most fantastic mammalian groups that ever existed. In the course of the Tertiary most of these species died out. Latter-day representatives of the *Notungulates* and *Litopternians*, as these South American ungulates are called today, probably lived on deep into the Pleistocene.

Darwin was thus the first naturalist to investigate the eccentricities of Tertiary mammalian fauna in South America, whose hothouse atmosphere had fostered the wildest of divergencies. South America's animal world during the Tertiary period looked like the doodlings of a surrealist artist; the palaeontologists often felt like Alice in Wonderland. For several epochs of the Tertiary the continent was completely isolated. During these intervals its animal species developed, as Wilhelm Bölsche has put it, 'the most singular offshoots; there arose a fauna gigantic in form, which today strikes us as a caricature of evolution'.

Along with curious creatures related to *Toxodon* and *Macrauchenia*, creatures partly recalling hares, partly pigs, tapirs, or elephants, partly horses or giraffes, there were subsequently found the strange *Astrapotheres*, which in reconstructions usually appear with melancholically drooping trunks, and the even more enigmatic *Pyrotheres*, whom the scientific artists portray with a pertly turned-up snub nose. Their classification is still in doubt; the *Pyrotheres* have not yet definitely been assigned to the ungulates and thus to the higher mammals.

On the other hand, Darwin discovered other fossils which could be properly compared with present-day species. The kinships were quite apparent. In the vicinity of the Argentine city of Rosario he came upon two enormous skeletons protruding from a cliff face on the shore of the Paraná. These were mastodons, representatives of a group of elephants that had also flourished in the Old World. There were also fossil horses, which particularly surprised Darwin, since it was not yet suspected that the equine stock had evolved in the New World. Fossil rodents from the clay strata of the pampas and the gravel of the riverbanks were remarkably like the typical South American rodents of today: the capybaras, vizcachas, and tuco-tucos.

Above all, however, Darwin was able to study a characteristic South American genus that seemed as strange and monstrous as the mysterious ungulates *Toxodon* and *Macrauchenia*: the gigantic relatives of present-day sloths, anteaters, and armadillos.

Giant Sloths in the Pampas

The hugest of the giant sloths that ever lived, the elephant-sized *Megatherium*, was even then the most 'popular' of fossil creatures. It was popular not only on account of its size, but on account of the involuntary role it had played in a scientific scandal. In 1789 a megatherium was first discovered in the cliff overlooking the Luján River, southwest of Buenos Aires. The bones were sent to Madrid, where the Spanish palaeontologist José Garriga had them carefully prepared and mounted.

The results were so strange that Garriga took six years to draw up a precise description of the colossal beast. Meanwhile, however, Cuvier in Paris had received through a French diplomat pictures of the skeleton and proof sheets of the Madrid drawings. Without bothering to clear the matter with his colleague Garriga, he published the find and assigned it its scientific name. The Spanish scholar felt that he had been shamefully cheated, for his detailed description, complete with five large plates, came out a year later. Cuvier, however, paid no attention to Garriga's protests, any more than he had acknowledged the complaints of Canon Godin, from whom he had stolen the famous mosasaur.

Cuvier was quick to see that the megatherium was related to the sloths. Somewhat later, in 1821, Eduard d'Alton, the German anatomist and engraver, published a magnificent volume of South American fauna in which he compared the megatherium with living sloths. Goethe referred to d'Alton when he wrote an essay designed to demonstrate that the tree-dwelling sloths of the Pleistocene must have descended from the extinct giant sloth.

As a matter of fact, Goethe was mistaken; for in reality the giant sloths were somewhat more closely related to anteaters than to present-day tree sloths. All three groups, however, different as they look, have common ancestors.

Whether Darwin had read Goethe's essay is uncertain. In any case, when he dug up all sorts of relatives of the megatherium at Bahía Blanca and along the Río Negro, he realized at once that these animals – some large as an elephant, some small as a tapir – were basically similar. He at once perceived the close relationships between the fossil giant sloths and present-day sloths and anteaters. Similarly, he promptly associated the extinct glyptodonts, which had been sheathed in tortoise-like armour, with living armadillos. But the question remained: did not these accumulations of fossil bones in South America offer an argument in favour of Cuvier's catastrophe theory?

In his travel journal Darwin noted: 'The number of the remains embedded in the grand estuary deposit which forms the Pampas and covers the granitic rocks of Banda Oriental, must be extraordinarily great. I believe a straight line drawn in any direction through the Pampas would cut through some skeleton or bones. ... The origin of such names as "the stream of the animal", "the hill of the giant", is obvious. ... We may conclude that the whole area of the Pampas is one wide sepulchre of these extinct gigantic quadrupeds'.

'Shake the Entire Framework of the Globe'

Darwin guessed right. Argentina turned out to be a paradise for palaeontologists. Only a few decades later scientists from several nations were digging there, including the German Hermann Burmeister and the Argentines Francisco Moreno and Florentino Ameghino. Everywhere in the Tertiary formations of the pampas they made discoveries. In addition to the curious ungulates, which had apparently existed nowhere else in the world, they came upon very ancient marsupials, homunculus-like primitive monkeys, and a number of other varieties of giant sloths and giant armadillos.

Why were such enormous quantities of mammalian fossils found in the pampas and in particular sites? Before Darwin, most geologists and palaeontologists would not have given the matter a second thought, but would have inferred a flood or some other catastrophe. But Darwin had Lyell in mind; he tried to deduce past events from present conditions. While he was in Patagonia, he witnessed the devastating effects of drought upon the livestock. Thousands of cows and horses died on the way to the watering places, or, reaching the rivers, perished because they had been so weakened that they could not climb back up the steep bank. The rushing waters of the next rainy season carried their bones into the pampas and buried them under the deposits of silt.

Could this not have happened in the past to the wild animals? 'What would be the opinion of a geologist,' Darwin reflected at his first sight of such heaps of bovine and equine bones, 'viewing such an enormous collection of bones, of all kinds of animals and of all ages, thus embedded in one thick earthy mass?' It would not occur to him that droughts or other natural circumstances were to blame; he would probably 'attribute it to a flood having swept over the surface of the land, rather than to the common order of things'.

The extinction of South American Tertiary fauna could also be misconstrued, Darwin decided.

The mind at first is irresistibly hurried into the belief of some great catastrophe; but thus to destroy animals, both large and small, in Southern Patagonia, in Brazil, on the Cordillera of Peru, in North America up to Behring's Straits, we must shake the entire framework of the globe. An examination, moreover, of the geology of La Plata and Patagonia, leads to the belief that all the features of the land result from slow and gradual changes.

In the course of the voyage he turned his mind repeatedly to the phenomenon of extinction. He was already on the track of a principal point of his later theory of selection when he remarked on the rapid reproduction of organisms and concluded that nature must have some means for checking too great an increase. Had unfavourable conditions of life made certain animals more and more rare, and finally brought about their disappearance? 'Certainly, no fact in the long history of the world is so startling as the wide and repeated exterminations of its inhabitants,' Darwin admitted.

If catastrophes were not to blame for the extinction of these species, there remained only one explanation as Darwin saw it: slowly, normally, as a result of natural events, in the course of time, the given species had become rarer and had finally vanished altogether.

To admit that species generally become rare before they become extinct – to feel no surprise at the comparative rarity of one species with another, and yet to call in some extraordinary agent and to marvel greatly when a species ceases to exist, appears to me much the same as to admit that sickness in the individual is the prelude to death – to feel no surprise at sickness – but when the sick man dies to wonder, and to believe that he died through violence.

Thus Darwin early in his career answered the catastrophe theorists.

Darwin thought it quite possible that some species, which he had found in a fossil state, might have died out only in historic times. The skull of a mylodon, a species of giant sloth about the size of an ox, was so slightly petrified that the animal might have been living a relatively short time ago. Mylodon remains in the cave of Ultima Esperanza, in the Chilean part of southern Patagonia, misled prominent palaeontologists into the belief that giant sloths had been kept as domestic animals by men, and had disappeared only a few hundred years before, if they were not still in existence somewhere. That was something of an exaggeration, but it is established that the mylodon (like other South American 'fossil' animals) was a contemporary of man. It had been hunted by Indians and smoked out of its caves.

Darwin gave deep thought to the relationships between extinct and living species. Could it be that some fossil forms had not really been extinguished? Had they been the true physical ancestors of present forms? He became more and more convinced that the present-day fauna was in many respects a retouched picture of the earlier ages: 'This wonderful relationship in the same continent between the dead and the living will, I do not doubt, hereafter throw more light on the appearance of organic beings on our earth, and their disappearance, than any other class of facts.'

When Darwin ended his world tour on 2 October 1836 he moved to London where he remained for a number of years, sifting and working over all the material he had brought back with him. By now he was no longer a self-taught beginner, but an experienced naturalist. The fossils he had collected went to the British Museum, where the most prominent palaeontologist in England, Sir Richard Owen, personally examined and described them. Since Sir Richard's reputation in England was equivalent to that of the recently deceased Cuvier in France, Darwin rightly felt it a special distinction that Owen should make use of his material. Darwin was now well known to contemporary scientists.

In later years Darwin and Owen were to be bitter opponents. But for the present they worked closely together, in a spirit of friendship. Richard Owen was one of those brilliant men whose start was inauspicious. In his schoolboy years, he was bluntly called 'lazy and impudent' by his teachers. But the cheeky lazy-bones developed into one of the leading scientists in England. Owen rose to the position of Superintendent of the Natural History Department of the British Museum, and was responsible for the creation of the Museum of Natural History in South Kensington, of which he became director. His contemporaries regarded him as infallible. Like Cuvier, he could classify a fossil simply by the structure of the teeth.

Owen studied the giant sloths, Darwin's favourite fossils, with exceptional care. He was able to dismiss a large number of erroneous ideas about them. D'Alton and Goethe had taken them for swamp or water dwellers, thanks to their peculiar physical structure. Other scientists assumed that they must have climbed about in trees, like present-day sloths. For proof they pointed to the creatures' large, sharply curved claws; an animal with such claws, they said, could not have moved along the ground. But if this were accepted, it became necessary to postulate – and so postulate they did – enormous primordial trees with branches strong enough to carry animals the size of elephants.

Darwin thought this idea nonsensical. He fully agreed with Owen's opinion that the huge animals had not climbed in trees, but on the contrary had drawn the branches down to them, or even uprooted trees, so that they could feed on the leaves. Knowing well the appearance of the bones, he was able to add:

The colossal breadth and weight of their hinder quarters, which can hardly be imagined without having been seen, become, on this view, of obvious service, instead of being an incumbrance: their apparent clumsiness disappears. With their great tails and their huge heels firmly fixed like a tripod on the ground, they could freely exert the full force of their most powerful arms and great claws.

As yet, he said not a word of any genealogic relationships between the giant sloths and present-day sloths. Darwin seemed to see eye to eye with Sir Richard Owen in all respects. In the beginning, the two men did not differ at all over Owen's special field, reptilian research. Owen's excellent monographs on the saurians of Lyme Regis and other fossil reptiles proved to be, on the contrary, a treasure trove for Darwin.

At the time there was no reason for the two to differ, for during his London years Darwin was still only gestating those ideas which would later cause such dissension. Darwin was no scientific rebel; he was a man who weighed all the empirical evidence, slowly groped his way forward, deferred publication. He was concerned to find out what the cause of evolution might be, and what factors could have governed it. He was not going to launch his theory before it was really weathertight.

In 1842 he withdrew to Kent, where he had bought a country property. There he lived quietly, pursuing his studies and his hobby of breeding pigeons. He had married a cousin, Emma Wedgwood, who tenderly nursed him through his frequent illnesses, for he had contracted a severe disease in South America. (It was probably Chagas' disease, which is transmitted by a tropical insect.) In time he published several essays on earthworms, the origin of coral reefs, and the biology of barnacles, all of which were received with respect by the scientific world.

But otherwise little was heard from Charles Darwin for fifteen years. Only a few friends had any idea that the invalid who lived in almost eccentric isolation was forging a theory that was to shake the entire world.

Struggle for Life

In 1798 a certain English rural vicar was struck by the thought that the Biblical commandment 'Be fruitful and multiply' was no longer apposite in changing times; that, in fact, considering the future of the human race, it was a highly dubious and dangerous injunction. He therefore wrote an interesting, bold, and to some extent revolutionary book on population policy. Since he did not know how his contemporaries would react to the breach of such a taboo, he cautiously brought the book out anonymously at first. In the next edition, however, he revealed his name: Thomas Robert Malthus. The book at once became the centre of a lively controversy, chiefly because Malthus warned his fellow men against continuing to bring so many children into the world. If the human race went on increasing in geometrical progression, while food supplies could only increase by arithmetical progression, he argued, the earth would soon be no longer able to feed the hordes of people. The inevitable consequence must be starvation and death for a large part of the population.

Malthus laid stress on the enormous over-production of living organisms, and pointed out that the descendants of a single pair of animals would populate the entire globe within a relatively short time, were it not for the struggle for life. Because of the strong competition, however, only a few individuals survived; thus Nature in her wisdom regulated the problem of increase. Man, however, being no longer exposed to this struggle, would be able to reproduce without bounds, unless he himself imposed limits upon his ability to propagate.

For a century and a half Malthus was regarded as a prophet by his disciples, the Malthusians and Neomalthusians. But the overwhelming majority of readers and critics dismissed him as a pessimist whose prophecies of gloom had long ago been disproved.

In the years to come the population of the globe continued to increase and new sources of food were found for man by putting virgin territories under cultivation. Thus – in the phrase of Edward Hyams, the English historian of culture – man became more and more a 'parasite of earth'. Today the alarming new population explosion demonstrates that Malthus was not altogether wrong. To be sure, his advice that the increase be stemmed by sexual abstinence was somewhat ingenuous, and is no longer pertinent; but in general Malthus correctly anticipated the present dilemma.

As early as 1838 Darwin came upon the phrase *struggle for life* in Malthus. For Darwin, this phrase became the key to his own theory. He saw that Malthus was right: the overwhelming majority of organisms are destroyed in the struggle for survival. But which ones survive? Darwin drew the conclusion, both patent and brilliant, that those best adapted to the given environmental conditions – the fittest – will live to pass on their useful characteristics to their descendants. In subsequent generations the same process of selection takes place. If this process continues over great spans of time, the degree of fitness increases constantly and the organism becomes better and better adapted to its environment.

If, however, external conditions change, other characteristics become requisite in the struggle for life. Other variants come to the fore, increase, and bequeath their useful characteristics to the next generations. Thus the appearance of the species slowly changes; a new form arises.

This theory could hold only if the offspring of a pair are not entirely alike, but incline constantly to variation (or as geneticists would say nowadays, if new mutations constantly arise). Here was something which could be proved by experiment. In order to build up evidence, Darwin raised a wide variety of animals and plants at his country place. He substituted artificial for natural selection, deliberately directing the selection process and testing the extent of variation and hence the effectiveness of the mechanism of selection.

Few scientists, certainly, have ever experimented, tested results, and gathered material for so long, with such perseverance, before venturing to publish the new knowledge thus obtained and the conclusions thus arrived at. Several of Darwin's friends, among them Lyell and the botanist Joseph Dalton Hooker, learned of his researches and ideas. After talks with the 'hermit' at Down they became half-persuaded of his views, and urged him to write a fundamental book about his theory. The idea was in the air, and if another and less thorough researcher anticipated the conscientious Darwin by presenting a shaky or incomplete hypothesis, the idea of evolution might be set back rather than advanced, and scientific progress impaired rather than furthered.

A Book Shakes the World

As it turned out, the impatience of Darwin's friends was fully justified. Someone actually did anticipate Darwin's theory. It was a young naturalist and globe-trotter who had been collecting animals in the islands of Indonesia, a 'man in storm and stress', as he called himself: Alfred Russel Wallace. Wallace too had observed that island-dwelling species of animals differ from the species on the nearby mainland; he too had read Lyell and Malthus, had concluded that geological changes would affect the forms of plants and animals; and had reasoned that excessive propagation necessarily involved selection of the fittest in a struggle for existence.

Wallace had heard something of Darwin's researches. He corresponded with his colleague and apparent rival in Down, and was more than delighted when Darwin asked him to set his ideas down in a formal treatise. The request prompted him to sketch out an essay which contained more intuitive insight than factual evidence. He sent it to England, hardly expecting it to create any stir in the scientific world.

Darwin, however, at once wanted to renounce his priority in Wallace's favour. Lyell and Hooker had great difficulty persuading him to put his own theory on paper in at least tentative form. At the 1 July 1859 meeting of the Linnean Society, the leading scientific organization in England, both papers were read. Oddly enough, the learned audience did not find the ideas of Darwin and Wallace especially world-shaking. They congratulated Darwin politely on his effort 'to explain the appearance and propagation of varieties and species on our planet', and encouraged him to work further on this problem. That was virtually the sole reaction of leading British scientists to a theory that in fact turned their own view of nature upside down.

This very lack of response gave Darwin the final spur to complete his work. He withdrew to the Isle of Wight and finished the book within a year, giving it the somewhat cumbersome title: *On the Origin of Species by Means of Natural Selection or the Preservation of Favoured Races in the Struggle for Life.*

The book was published in November 1859. It caught on at once. In spite of the austere scientific content, the first edition sold out on the day of publication. Friends and foes seemed to guess that it would usher in a new scientific age and give rise to a profound intellectual revolution.

Hitherto most naturalists had believed in a rigid, immutable nature and a transcendental creation. Darwin abandoned this concept. For Darwin, nature had no predetermined aims in mind; it constantly produced a seemingly purposeless abundance of forms and variants. The best fitted won through and passed on the torch of life; the unfit lost out in the struggle for life and were extinguished. Man, too, was subject to this process. He could not be set apart from the world of all other organisms as 'crown of Creation' or 'measure of all things'; he sprang from the animal kingdom and had developed by the same laws as other organisms.

These ideas and conclusions had the effect of a shock upon many persons, inspired swift and enthusiastic agreement in others. In England, particularly, the soil was favourable to this outlook. Adam Smith, the father of English liberalism, had called free competition the motive force of human development. The Manchesterism of the Victorian era stood for the greatest possible freedom in commerce and economic life. Free competition was promptly identified with the Darwinian struggle for life; Darwinism was seen as an appropriate scientific codicil to the theories of Adam Smith. From England, Darwin's theory fairly easily conquered liberal circles elsewhere, first in Germany, then in the rest of Europe and in America.

The Darwinian doctrine seemed to appeal to men of the most diversified philosophical views. Thus, the positive philosophers of the nineteenth century were almost without exception Darwinists. Certain ducal houses whose princes were receptive to enlightened and freethinking ideas took up the doctrine. Liberal groups in the Protestant Churches wondered whether the idea of evolution might not be more compatible with the Christian doctrine of Creation than the prevailing catastrophe theories. As for the pantheists – a sizable contingent in the nineteenth century – they found much that was highly congenial in Darwinism.

Yet the same scientific theory that commended itself to the liberals because it served as a biological justification of free enterprise also suited the socialists, providing as it did a biological basis for the theory of class struggle and belief in steady historical progress. In time the communists, too, turned to Darwin, although they rejected the sustaining pillar of his doctrine, selection. Actually, Lamarck's milieu theory was more in tune with their political ideas – in spite of all the friendly words they have cast Darwin's way since the time of Marx and Engels.

Thus the most heterogeneous groups tried to use Darwin's theory for their own ends. Even nationalists and colonialists could invoke the theory; they gave a sinister twist to Darwin's principle of the 'survival of the fittest'. If it was the law of nature that only the fittest were destined to survive, then wars of conquest, colonial campaigns, and claims by the presumably fitter white 'master race' to dominate the presumably 'lower' races were obviously legitimate.

As for Darwin himself, he was most seriously affected by all this. He became, against his will, one of the most controversial personages of his age, indeed of modern times. Perhaps not since the Inquisition's trial of Galileo had any scientist been so misunderstood, misinterpreted, slandered, and made the whipping boy for the most extreme philosophical, political, economic, sociological, and universally human vagaries as Charles Darwin.

'Red in Tooth and Claw'

The misunderstandings began in Darwin's lifetime. They were so unpleasant that Darwin soon refused to take part in scientific disputations. It distressed him to see opponents growling at one another 'in a manner anything but gentlemanlike'.

The quarrels flared up particularly over the old concept borrowed from Malthus, the 'struggle for life'. Darwin's enemies painted him as the spokesman for crude utilitarianism, lust for profit, ruthless egotism, aggressive war, and similar brutal conduct. As time went on, it was said that Darwin's 'struggle for life' provided an excuse for imperialism, pan-Germanism, fascism, racialism, and even the genocide of the recent past.

But what did Darwin really mean by the often-quoted 'struggle for life', which – it was claimed – taught men the law of the jungle, the ethics of elbowing their way through life, and other such pernicious lessons? Did Darwin see nature as really so cruel, so 'red in tooth and claw', as Tennyson put it? The very personality of Darwin argues against any such interpretation.

Darwin belongs amongst the most humanitarian thinkers in history. He was well disposed towards people, had a great affection for animals. From his voyage around the world he brought back shocking accounts of the misery of the slaves, and ever afterwards spoke out for the weak and the persecuted. On his rides through the desert of northern Chile, he suffered pangs of conscience if he were unable to feed and water his horses on schedule. He could not bring himself to kill animals. He shielded every living thing he could from harm. It is hard to believe that this extraordinarily soft-hearted man regarded nature as an unending war in which every creature savagely attempts to annihilate every other.

Nor did he. For Darwin the struggle for life was not a bloody contest of individuals or species against one another. Rather, the phrase connoted a constant struggle for existence to which different varieties and populations *within a species* are exposed. The varieties (or as we would say today, the mutations) of a species do not 'struggle' against one another; they are merely subject to a constant testing; and the fittest, those individuals that carry the most useful mutations, which can best adapt to environmental conditions, win out in this test.

'It may metaphorically be said that natural selection is daily and hourly scrutinizing, throughout the world, the slightest variation; rejecting those that are bad, preserving and adding up all that are good; silently and insensibly working, *whenever and*

The most famous fossil: the Solnhofen archaeopteryx, a 'feathered lizard'. The controversy waged for a long time: bird or reptile?

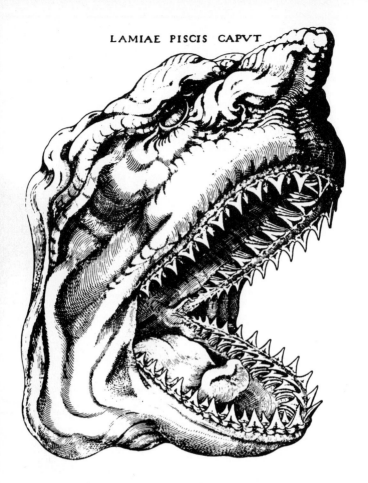

The head of the primeval shark as Steno pictured it. *(From Nicolaus Steno's* Prodromus, *1667.)*

Scheuchzer's 'witness of the Deluge' was in reality a giant salamander of the Tertiary.

The forgeries
that deceived
Professor Beringer
of Wurzburg.

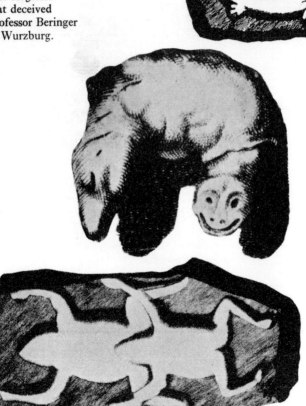

The illustrations in old books reveal a touching effort to depict life in the primordial world. (*Above*) A nineteenth-century effort representing the **saurians** of the Lias ocean.

(*Above*) Saurians of the Cretaceous in combat. Only in recent decades have reconstructions approached more closely to the probable reality.

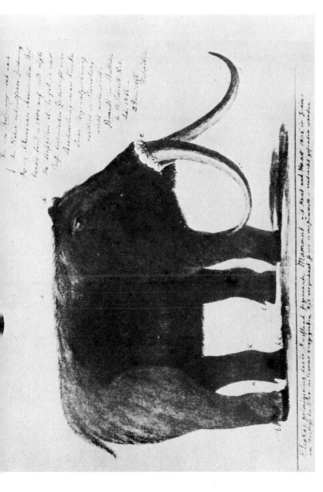

Ideas of how the ancient animals had looked underwent considerable change as science penetrated more deeply into the past. (*Above*) The famous drawing of a mammoth by Boltunov, the Russian ivory trader, done in 1806.

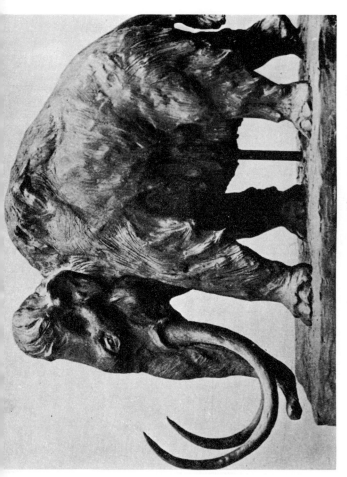

(*Above*) Modern reconstruction of a mammoth by palaeontologists Othenio Abel and Franz Roubal.

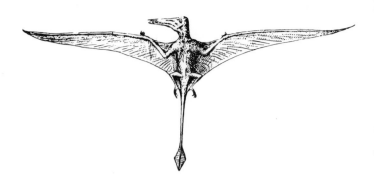

When the first flying lizards were found in the slate quarries of Solnhofen, the most contradictory theories were propounded on the habits of these strange animals. Only Cuvier saw the answer to the riddle. *(After a drawing by Othenio Abel.)*

The gigantic cave bear, more vegetarian than predator. The reasons for its extinction have been studied with particular care. *(From a drawing by Othenio Abel.)*

In Patagonia, Darwin found one of the most curious animals that ever lived in the New World, *Macrauchenia patachonica*.

Giant sloths of the South American Tertiary. Their kinship with present day sloths, anteaters, and armadillos prompted Darwin to reflection on the origin of the species.

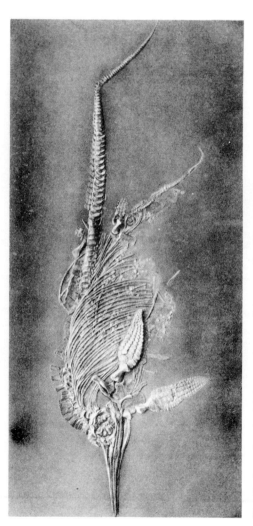

Fossils show that the ichthyosaurs were viviparous. The female in this photograph was killed by some calamity during the parturition.

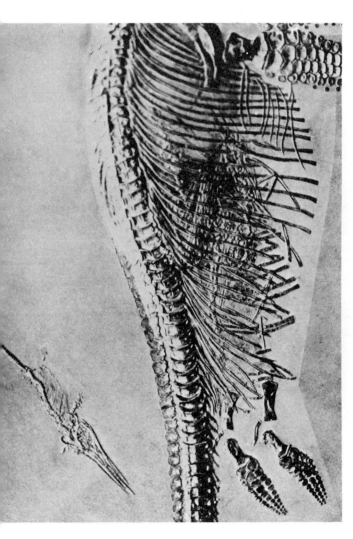

In the female shown here, along with the newly born young animal the embryos can still be seen in the mother's abdomen.

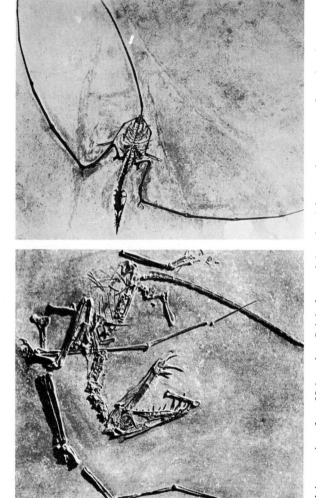

Flying saurians from Holzmaden, Solnhofen, and America (*above and next page*); from the pigeon-sized saurian to the mighty king of the air which was bigger than any bird now living.

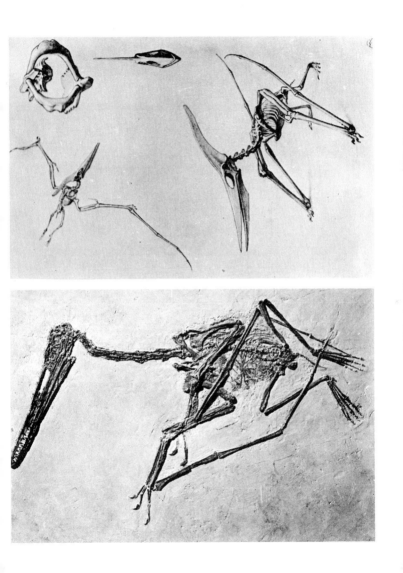

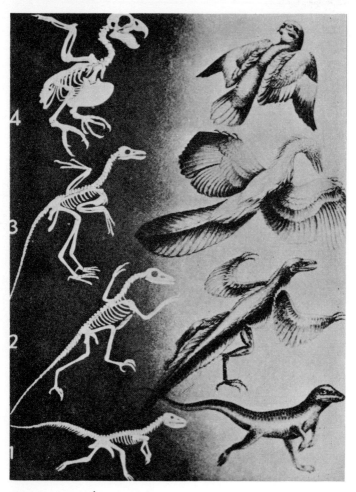

Huxley proposed the theory that birds are reptiles in disguise.
(From bottom to top) Bird saurian, avian ancestor, primordial **bird**,
bird.

wherever opportunity offers, at the improvement of each organic being.'

The notion that Darwin's theory held living organisms to be the products of blind chance is equally without basis. Darwin suspected what geneticists have by now verified in countless experiments; that as soon as a new situation arises in the environment, as soon as prospects for a new kind of life open up for an animal or a plant species, a period of lively mutation begins. 'The organism adapts to the situation by producing a suitable stock if there is any such potentiality within its hereditary structure,' as the German biologist Gerhard Heberer has put it.

The mechanism of evolution is not a blinding churning of chance; it takes place in a statistically comprehensible manner. Slowly, imperceptibly, but in keeping with fixed rules, a new species evolves. At first the members of a species begin to mutate. These mutations are as yet not directed towards any goal; genetic variety is what increases the possibilities. 'Nature does not aim, she plays,' Otto Renner, the German botanist, has commented.

The factor of natural selection begins to operate; it is selection that introduces a goal into the evolutionary process. The outcome depends on the environmental conditions. If the genetic improvements increase, the adaptations to environment grow more and more perfect. Finally the species has changed so fundamentally that it can prosper under the new conditions of life.

The Austrian naturalist Konrad Lorenz, making what amounted to a profession of faith in Darwin, has had this to say about the theory of selection and what it means to us, the human race:

'It possesses all the qualities a cosmogony can have: persuasiveness, poetic beauty, and impressive grandeur. Once we have perceived this, there is nothing repugnant in Darwin's insight that even he himself owes his existence to this most magnificent of all natural phenomena, and that we are of one stock with the animals of this earth – any more than there is in Freud's perception that we too are driven by the same instincts as our prehuman ancestors. We may perhaps feel a new kind of awed reverence – reverence for the achievements of reason and moral feeling, which first came into the world with the appearance of man, and which may well give man the power to control his animal heritage, so long as he does not, in blind arrogance, deny its existence within himself.'

Darwin and his followers drew this very conclusion from their new knowledge. The theory of evolution, as the German biologist Ernst Haeckel put it only two years after the publication of *The Origin of Species*, would contribute 'more than anything else not only to the intellectual liberation but also to the moral perfection of man'. But on this point there was much disagreement.

'We've Always Known That'

For palaeontology, Darwin's theory initially demanded such radical reconsiderations that many workers in the field long hesitated to accept it. They would have to throw all their books into the fire if they did. If the theory of evolution were correct, generations of scientists had been chasing will-o'-the-wisps. That was a difficult admission to make.

Shortly after the publication of Darwin's epoch-making book, Wilhelm Keferstein, the famous Göttingen zoologist and a fierce enemy of evolution, could see what it portended for his science: 'For centuries competent and ambitious scholars have done fine work in the service of zoology, and have constructed excellent systems. They have succeeded in shaping a general opinion on their subject in the mind of humanity. Now a new man comes along who wishes to topple the structure reared with so much toil.'

Darwin himself had pointed out to the palaeontologists the faultiness of the record. 'The crust of the earth with its imbedded remains must not be looked at as a well-filled museum, but as a poor collection made at hazard and at rare intervals.' Finds of fossils are a matter of good luck; Darwin therefore seriously doubted whether it would ever be possible to reconstruct the whole history of life from petrified remains. Nevertheless, he thought, the palaeontologist must be on the lookout for the intermediate links (what later would be called the missing links) – for 'the number of intermediate varieties, which have formerly existed [must] be truly enormous'.

Darwin seemed to be directing palaeontologists to work as genealogists. 'In the future I see open fields for far more important researches,' he prophesied. 'Much light will be thrown on the origin of man and his history.'

The inclusion of man in his theory brought Darwin into sharp conflict with orthodox theologians and all conservative forces in his country and elsewhere. Many scientists of religious disposition, including the two leading palaeontologists of the age, Louis Agassiz and Richard Owen, felt that the theory of evolution would have an ill effect on the moral development of man. They threw the whole weight of their personalities and their immense scientific prestige on the side of Darwin's opponents. The first reaction of a religious conservative to Darwin's revelations was summed up by Prime Minister Benjamin Disraeli in an address to the students of Oxford as early as 1864. The question facing society, Disraeli declared, was simply: 'Is man an ape, or an angel?' And his own answer was 'My God, I am on the side of the angels.'

'It is Like Confessing a Murder'

Disraeli was putting the matter in somewhat extreme terms when he phrased the alternative: 'Ape or angel.' But naturalists did realize quite clearly that the real alternative was: creation or evolution. Not only the Biblical narrative of the Creation, not only the uniqueness of man, but also the immortality of the soul and the question of God's existence were involved. Owen and Agassiz perceived that at once; but Charles Darwin, erstwhile student of theology, also saw the bearing of his theory. He understood the resultant dilemma perhaps better than many of his opponents.

Unlike the Puritan Owen and the Calvinist Agassiz, Darwin was at home in a liberal form of Christianity; hence he found it easier to free himself gradually from religious dogmas. He probably shared the views of his friend Herbert Spencer, the philosopher, who held that inadequate man could know nothing about God and must therefore exclude him from philosophical considerations. Darwin, as Gerhard Heberer puts it in a recent biography, 'was a man who had a sense of the great mystery which is given to us with our existence, but was too modest and too reverent to venture any human statements about this mystery'.

A person of this sort inevitably felt inner conflict when his researches led him to facts and principles which seemed to argue against the tenets of religion. Perhaps that was why Darwin had been so hesitant about announcing his theory – far more hesitant, at any rate, than the majority of his followers. He knew that his theory might lead to a philosophical and theological upheaval. That parenthetical remark of his in his letter to Hooker when he realized that species were not immutable – 'it is like confessing a murder' – indicates that he was himself just as shaken as his contemporaries by the implications of his theory.

Some biographers contend that the severe stomach ailment which plagued him until the end of his life was psychosomatic. The constant religious conflicts, they reason, generated enough 'stress' within him to cause irreparable physical harm. This fashionable explanation may be wrong; more likely Darwin's chronic illness was a repercussion of the Chagas' disease he picked up in South America.

Be that as it may, it is clear that Darwin was confronted with the very same problems that still confront every Christian: how can our scientific view of the world be reconciled with the Christian view? What becomes of Christianity if it must accept the scientific facts? What becomes of the sciences if they continue to hold to Christian dogmas? These problems have been troublesome since the days of Giordano Bruno, Copernicus, and Galileo; but in Darwin's era the world became more conscious of them than ever before. And Darwin himself fought out the resultant conflict within himself.

When the world-wide battle over Darwin's theory finally erupted, therefore, the reasons behind the fervour of many important scientists were as much religious as scientific, if not more so. The replacement of teleology by causality, of creation by evolution, and the abolition of man's special role by including him in the course of nature, was so revolutionary that even the foremost minds of the age were profoundly disturbed. Members of Darwin's own family were troubled. In 1871, when Darwin at last began dealing with these dangerous implications of his theory, in a work to be called *The Descent of Man*, his wife Emma commented that it was certainly an interesting and important subject, but it was pushing Darwin still farther from God.

The scientific arguments against Darwin were not nearly so

weighty. In fact, they sounded distinctly weak. Anti-Darwinist palaeontologists could only put forward the views of Cuvier, which by now had suffered considerable battering. Darwin, in fact, at first laboured under the delusion that he had found brothers in spirit in his two chief opponents, Owen and Agassiz.

Richard Owen liked to refer to the 'everlasting activity of a creative force' which in some indefinite and indefinable manner saw to the orderly creation of living organisms. Moreover, in all his publications Sir Richard had constantly stressed the relationship between fossil and living species. On the basis of these things Darwin had assumed that Owen was among the palaeontologists who were completely convinced of the mutability of species.

The case of Agassiz was similar. The man who had discovered the Ice Age, the world's greatest authority on fossil fishes, had meanwhile acquired American citizenship and was teaching zoology at Harvard University. So great was his popularity that his lectures appeared in the New York *Tribune*, and enjoyed a far wider readership than was granted to most professors. He and his associates and pupils managed to arouse enormous enthusiasm for the sciences in the United States. Agassiz, thanks to his charming personality and powers of suggestion, also had a faculty for obtaining sizable contributions for research from millionaires – among them people who previously 'would sooner have died than give a cent to Harvard University'. Agassiz converted them into generous patrons, induced them to set up foundations, and thus established the tradition for foundation support of the sciences in the United States.

In his major work, *Les Poissons fossiles*, Agassiz advanced several theses which implied a certain sympathy with the idea of evolution. For example, he pointed out that primitive fossil fishes resemble the embryos of present-day species of fishes – thus anticipating the biogenetic principle formulated by Darwin's supporter Ernst Haeckel. 'Agassiz and several other highly competent judges insist . . . that the geological succession of extinct forms is nearly parallel with the embryological development of existing forms,' Darwin wrote. 'This view accords admirably well with our theory.'

Equally exciting were Agassiz's many references to 'connect-ing links': the wormlike lancet fish, the salamander-like lung-fishes, the egg-laying duckbill platypus, and other species which combine the characteristics of several groups of animals. To be sure, Agassiz forbore to conclude from such evidence that species were mutable; but Darwin was forced to assume that 'such forms are in fact transitional or connecting links'.

Thus the observations of both Owen and Agassiz struck Darwin as confirmations of his own theory. It did not occur to him that these two men, whom he esteemed so highly, would launch the most furious attacks against him. Nevertheless, Agassiz became head of the anti-Darwinists in America. In lectures and publications he constantly vociferated against the 'wanton theory' and the 'sensational zeal' of Darwin's followers. He regarded the 'new scientific doctrines that are now flourish-ing in England' as fads and prophesied that they would soon be outmoded. When asked what evidence he had to refute Darwin, he replied: 'All the facts of natural history proclaim the One God Whom we know, worship, and love.' Science, he maintained, had only one task: to analyse the thoughts of the Creator of worlds.

Owen did not speak of the Creator of worlds, but he volun-teered his services as a scientist to help the orthodox clerics of England in their war against Darwinism. His authority was as great as that of Agassiz, and the public was much influenced by the opinions of these two men.

Darwin's Bulldog

Owen was soon to meet with his first defeat. On 30 June 1860 a meeting of the British Association for the Advancement of Science was held at the Museum of Oxford University. Prom-inent scientists and other persons in public life gathered to dis-cuss Darwin's theory. Darwin himself did not take part; he abhorred such disputes. In his stead Thomas Henry Huxley, who had first reviewed *The Origin of Species* in *The Times*, under-took to explain the theory of selection, the descent of man, and the similarities in structure of human and ape brains.

Spokesman for the other side was Samuel Wilberforce, the

Bishop of Oxford, who because of his unctuous manner bore the nickname 'Soapy Sam'. Wilberforce had gone to Owen for briefing on the scientific questions. But he was driven farther and farther into a corner by Huxley. At last he could only try to ridicule Darwinism by an insulting joke: 'If you truly believe that you are descended from an ape,' he challenged Huxley, 'it would interest me to know whether the ape in question was on your grandfather's or your grandmother's side.'

Huxley's reply had been handed down in various versions. But the substance was: 'If you ask me whether I would prefer to have a wretched ape for my grandfather or a brilliant man of great importance and influence who uses his gifts to make mock of a serious scientific discussion, then I unhesitatingly declare that I prefer the ape.'

With these words, Huxley had seized the initiative. Instead of the shy and reticent Darwin, Sir Richard Owen henceforth was pitted against an opponent who could defeat him with his own weapons.

Thomas Henry Huxley was one of the foremost vertebrate palaeontologists in England. The family trees he sketched out were brilliantly confirmed by later finds and researches. On a world tour that in many respects resembled Darwin's he also developed into a highly qualified oceanographer. His chief interest, however, was the origin of man. Only four years after the publication of Darwin's theory of selection he delivered three sensational lectures on *Man's Place in Nature*, in which he described the ancient men discovered up to that time, above all Neanderthal man, and compared them with various lower and higher species of apes.

Huxley had the temperament, the boldness, and above all the requisite knowledge for the fray before him. He called himself 'Darwin's bulldog'. In general he had a penchant for canine images; after the publication of *The Origin of Species* he wrote to Darwin: 'As to the curs which will bark and yelp . . . you must recollect that some of your friends at any rate are endowed with an amount of combativeness that will stand you in good stead.'

Huxley lit into the doctrine of 'archetypes', which Owen attempted to oppose to the theory of evolution. Sir Richard had proclaimed that the many apparent concords and transitional forms in nature were due not to evolution, but to an archetypal plan of Creation. Every type, he maintained, must be the embodied form of an idea, and as such fixed and immutable. This explanation, he held, was at once more natural and objective than the thesis that all later forms derived from earlier varieties.

The theory Agassiz opposed to Darwinism had a similar cast, but was even more strongly founded upon religion. Agassiz believed in a personal God who populated the earth with ever more perfected beings in each epoch. No species, no animal form, no group of organisms derives directly from any other. Although each branch is part of a larger systematic unity, each is 'independent in its physical existence'.

Agassiz at times took the same view as Owen: 'The palaeontological succession of lower and higher beings is nothing but the gradual realization of the divine plan, which proceeds from the lower to the higher.'

The school of Agassiz took, on the whole, a simplistic view of the problem. Whenever they could not explain certain phenomena that seemed to indicate evolution, they invoked God, exclaiming that the Creator must have known why he had arranged things in one way and not another. Huxley was not far wrong when he pointed out that such an attitude was tantamount to blasphemy. Or as the German Darwinist Carus Sterne put it, it was 'an intense anthropomorphization of the divine creator, whom a zoologist created in his own image'. If the appeal were going to be to God, then there was no need to postulate abstract archetypes and phases which could not be reconciled with the facts of biology; better to admit at once that God had created organic nature with the aid of evolution and selection.

Many naturalists who at first took anti-Darwinist stands for religious or philosophical reasons gradually came over to Darwin's camp. They realized that even from the Christian standpoint the opposing theories were too artificial and hence unsatisfactory, that it was quite possible to believe in evolution without denying God. Although the conflict with Darwinism and evolution smoulders on to this day, in spite of many a compromise and peace treaty, the romantic, idealistic theories of creation, which burgeoned in such numbers during the decades of struggle, have long since fallen into oblivion.

It proved fortunate for Darwin and Huxley that the old idea of creation was once more taken up, explored, and pursued to its ultimate consequences by the best minds of the age. For the process itself exposed the amount of conjecture and the number of contradictions in which even excellent scientists necessarily became entangled as soon as they ceased to carry on their research without bias, but instead trimmed their results to preconceived opinions. Huxley need not have taken so much trouble to refute the opposing arguments; they as good as refuted themselves.

Owen and Agassiz remained anti-Darwinists to the end of their lives. Most palaeontologists, however, eventually came around. So, even, did Agassiz's son and successor, Alexander, who after some demur also accepted evolution, remarking ironically that whenever a new and astonishing scientific truth is discovered, people first say: 'It isn't true,' then, 'It's against religion,' and finally, 'We've known that always.'

With Darwin, a new age of great palaeontological activity began. Many of the finds – primitive reptiles from all parts of the globe, primitive birds from the Solnhofen slate, primitive horses from North American alluvial deposits, primitive giraffes from Greek gravels, primitive apes from the European Tertiary strata, or primitive men from the caves and clays of three continents – again and again confirmed Darwin's ideas. Huxley was prompted to exclaim: 'If the theory of evolution had not existed before, palaeontology would now have to posit it.'

In the Jurassic Sea

The mountains of Swabia in south-western Germany became more and more a centre of German palaeontological research. A veritable guild of palaeontologists had formed there, and when word of Darwin's theory reached the region it was received by men who diligently collected their ammonites, belemnites, and saurians, sometimes entertained bold ideas, but always kept to the certain ground of facts. Among them were pastors and other pious-minded folk. It should have been hard for the idea of evolution to make a breach in this tradition-bound, sometimes rather eccentric circle of scientists.

But curiously enough, the scientists of Swabia had been on familiar terms with the idea of universal evolution long before Darwin aired his theory. The mentor of the Swabian palaeontologists was Friedrich August Quenstedt, who taught geology in Tübingen from 1837 to the end of his life. Quenstedt made Tübingen a palaeontological centre. On a purely empirical basis, uninfluenced by other scientists or theorists, he also had arrived at the conviction that the geological epochs and their creatures were linked with one another by 'innumerable transitions'.

None of the Swabian palaeontologists found this statement objectionable. They nodded thoughtfully when they came upon such sentences in Quenstedt's *Epochs of Nature* as: 'Everything is in process; scarcely has the developing thing reached its peak than it must descend again, yielding place to others. ... Such a mighty rising and sinking in nature naturally can induce changes in animals and plants, changes of which we cannot know the full extent. ...' It was only after the general hubbub over *The Origin of Species* had begun that they too found the idea of evolution a subject for argument. Complaints were voiced that the naturalists were now engaged in the regrettable enterprise of eliminating miracles from the world.

Quenstedt took another view. 'We maintain that the miracle is still there.' Evolution to him was not equivalent to cold materialism, but quite as wonderful and awe-inspiring as the former view of Creation. His numerous disciples were much of the same mind.

Quenstedt focused his studies on the characteristic animals of the Jurassic in Swabia, the ichthyosaurs. An arm of the sea had covered part of the area during the Upper Jurassic, he concluded. The sedimentation at the bottom of this bay was highly stratified. Quenstedt used Greek letters to distinguish the Liassic stratifications. The ichthyosaurs in a number of the famous sites of Württemberg were mostly found in the layer he labelled as Lias epsilon.

In an age when most palaeontologists still believed in successive geologic cataclysms, Quenstedt carried out a detailed study of ichthyosaurian fossils which proved that the animals must have died natural deaths:

The fish-saurians died and sank into the ooze, which kept their undersides intact. Only the upper surface was subject to wear and tear. Slight waves washed it; the scaly shell separated from the skeleton. But the water was not strong enough to carry the detached parts very far; they were covered almost on the same spot by successive deposits of sediment. Thus neither the storms of the sea nor the savage power of mighty cataclysms killed these creatures. They died a natural death.

Fin and Claw

A pupil of Quenstedt was investigating the Lias epsilon of Bad Boll and other fossil-rich sites of Swabia. He was the familiar combination of parson and geologist: Oskar Fraas, the pastor of Laufen on der Eyach. Fraas had taken up the study of fossils out of an amateur's passion for such matters, but he was also helping his impoverished parish. Profits from the fossil trade went into the welfare fund. Many of the inhabitants of Laufen owed their rescue from destitution to Pastor Fraas and his ichthyosaurs.

In his book *Before the Deluge*, Fraas drew a vivid picture of the mass graves of the ichthyosaurs: 'There they lie in their stone coffins, wrapped closely round by slate; only the rough outlines can be distinguished, like mummies in their shrouds. The head is lightly defined, the spinal column, the position of the extremities, the whole length of the animal; and at a glance the workman recognizes by the shape whether it is an animal with fins or claws.' The quarry workers called the ichthyosaurs 'finned animals', and the rare ancient crocodiles 'clawed animals'. According to Fraas, the clawed animals were worth three times as much as the finned animals.

'But the price is not solely governed by that,' Fraas continues.

The most important factor is how and where the animal lies, and whether there is nothing missing, whether the slab is cracked in two. ... The workman does not make any effort to sell the find; he quietly puts it on display, for he knows that almost every week the buyers come, suppliers for curiosity cabinets and scientific collections. No horse-trading is carried out with such zest, with such outlay of eloquence and application of all sorts of tricks and arts, as saurian dealing. No commerce requires so much cunning along with precise knowledge of the commodity, if the buyer is not to suffer loss – since he is in any case buying a pig in a poke. Finally no deal is closed without the buyer's undertaking, as a special obligation, to hold a wake for the saurian with bottles of wine and cider.

From the title of his book it is evident that Fraas was a pious Bible believer who found it a pity that palaeontologists had to abandon the fine old legend of the Deluge. For a while he resisted the new ideas, but he did not fight against them; he knew too much about fossils to do so. At the age of thirty-three he finally abandoned the ministry for geology. Ultimately he became curator of the Stuttgart natural history collection; he proved himself a master in the preparation of fossils, a technique that to this day has remained exceedingly delicate work, calling for magnifying glass, microscope, penknives, and a variety of precision instruments. At the time of Oskar Fraas the procedure was as follows:

Now the time has come to 'clean' the saurian, that is to free it from its slate wrappings and bring its old bones into the light of the sun. Such work can be left only to initiates; an ignorant hand will ruin the animal. Many take months to do a single fossil; for the stone must be removed from the bone more with a graving tool and needle than with hammer and chisel. Anyone who has not himself employed the graving tool knows nothing of the joys that fill the heart of the connoisseur when he follows the course of a bone in the slate, exposing a bit more of it every day, until at last the harmonious whole of the animal lies before his eyes.

The Enigma of Holzmaden

Fraas was an early frequenter of the slate quarry in the tiny village of Holzmaden, which turned out to be the most important and richest ichthyosaur site on the face of the earth. The quarry had been bought by a chemist named Hauff, another former parson, who meant to start a plant to distil oil from slate by carbonization. But competition from the new petroleum industry forced Hauff to shift to the manufacture of slate table tops, stove underlays, and siding. It was a modest business involving heavy physical work, and offered not much more than a livelihood.

One of his sons, Bernhard Hauff, was not content with being a quarryman. He had many a talk with Quenstedt, Fraas, and the other scientists who came to Holzmaden looking for fossils. The seventeen-year-old showed such intelligent interest that Fraas at last proposed to the parents that he take the boy under his wing in Stuttgart and teach him the art of preparing fossils. As it happened, the boy spent only half a year with Fraas as an apprentice in palaeontology; then his father needed him in the family business once more. But Bernard Hauff had such extraordinary scientific and artistic gifts that with only this brief training he became a palaeontologist with a world-wide reputation. The ichthyosaurs he discovered, along with numerous other fossils, incidentally saved the business and finally relieved the Hauff family from any financial difficulties.

About a hundred ichthyosaurs a year have been found in Holzmaden ever since – although most have been so incomplete as not to be worth extracting from the slate. Hauff sold the better specimens to museums all over the world. Most of the European Liassic ichthyosaurs in collections today came from Holzmaden. The appearance and life habits of these characteristic animals of the Jurassic seas were finally clarified on the basis of specimens found there.

In Cuvier's time it was not yet known that the fish-dragons

had a shark-like tail fin. Consequently, the first reconstruction drawings endowed them with a long, lizard-like tail. In addition, ichthyosaurs were thought to come out on land from time to time, like seals and crocodiles. In fact, it turned out that the animal had been entirely a creature of the open sea. The shark's tail fin was attached to a body with the torpedo shape of the dolphin. Hence, ichthyosaurs could not have climbed out on land. They lived, as became obvious from the stomach contents of later finds, chiefly on belemnites. But how had they reproduced? Egg-laying reptiles, such as marine tortoises, for example, must normally crawl out on dry land in order to bury their eggs in some sun-warmed brooding site. Did this mean that the ichthyosaurs brought their young into the world alive, like whales and dolphins?

Old, erroneous representation of an ichthyosaur, from the middle of the nineteenth century.

For decades Hauff attempted to prove that this had been the case. Forty years previously Quenstedt had described ichthyosaur skeletons which seemed to contain the bones of young individuals of the same species. 'At times they seem to have devoured their young,' Quenstedt had concluded, but added: 'Or are these embryos in the mother's womb?' Later, more instances cropped up in which the remains of young were found in the abdominal cavities of mature individuals. Hauff alone collected fourteen such specimens which might be those of either cannibalistic adults or pregnant females.

Most experts were inclined to believe that the fish-dragons had been cannibalistic rather than viviparous. There were a number of indications. For example, the bones of some young saurians showed injuries which could only have come from the teeth of predatory adults. The state of decay of other young saurian remains in the abdomens of mature saurians had advanced to a point that suggested the processes of digestion. There was good reason to think that ichthyosaurs had occasionally eaten their own brood.

But had that been so with all such finds? Hauff doubted it, and did not give up his search for saurian embryos in the mother's womb. At last he found what he was after: his quarry yielded a slab of slate that silenced all arguments. It showed a splendidly preserved female with three embryos in her body, and on the point of giving birth to a fourth.

How had this saurian mother been surprised by death in the act of parturition? Why had the onetime Liassic sea near Holzmaden taken the lives of so many ichthyosaurs of all sizes and ages? When many thousands of animals are found dead in a confined space, it necessarily seems to imply some cataclysm. During Liassic time the bay of Holzmaden presumably was not filled with gaily disporting life; it seems to have been nothing but a gigantic ichthyosaurian cemetery.

Some scientists reason that the ichthyosaurs died in the open sea and were carried by steady currents into the bay. There, in the course of ages, innumerable cadavers sank to the bottom and were covered by sludge. According to another view, the saprogenic sludge in the bay developed hydrogen sulphide which poisoned the whole bay. When ichthyosaurs pursuing squid or escaping from enemies happened into this bay, they were quickly poisoned by the water, and since a sand bar largely blocked the entrance, they could not escape in time.

But neither theory is quite satisfactory. The number of ichthyosaurs in the Liassic strata of Holzmaden is too huge; it seems inconceivable that drifted cadavers or straying individuals should have accumulated in such quantities. The mystery remains.

A supposed ancestor of the birds: the ornithosuchus.

It was decades before palaeontologists realized that *Dinotherium giganteum* was an elephantlike animal. *(Reconstruction after Othenio Abel.)*

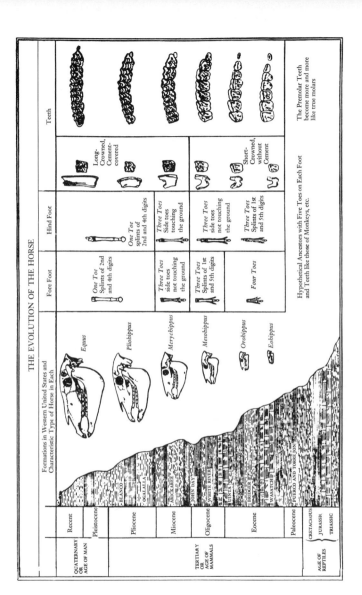

THE EVOLUTION OF THE HORSE

The duck-beaked trachodon was the first dinosaur whose discovery in North America created a great sensation.

Brontotheres, Marsh's great discovery during the Sioux wars.

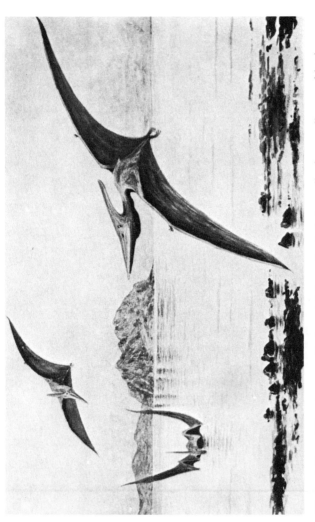

Giant relatives of the pterosaur, the largest flying animals that ever lived, were discovered in the American Cretaceous in 1870.

The subject of much astonishment and verse – the stegosaur with his two brains. 'No problem bothered him a bit, he made both head and tail of it,' Burt L. Taylor wrote in a famous snatch of doggerel.

Horned armoured dinosaurs, a late and curious branch of the saurians.

The American triceratops, the monster with enormously formidable defensive weapons.

Skull of the ancient whale *Zeuglodon*, the fossil from which Albert Koch constructed his notorious sea serpent.

From this detailed drawing of a titanothere jawbone, Osborn arrived at the idea that the animals might have become extinct because their teeth could not adapt to grass-eating.

The stegosaur. *(From a drawing by Othenio Abel.)*

Marsh crowned his lifework with the discovery of such gigantic saurians as these in the *Atlantosaurus* strata of Wyoming. *(After a reconstruction drawn by Othenio Abel.)*

The footprint of a dinosaur creates a tub holding seventeen gallons **of** water.

In the Gobi Desert, Roy Chapman Andrews came upon these dinosaur eggs.

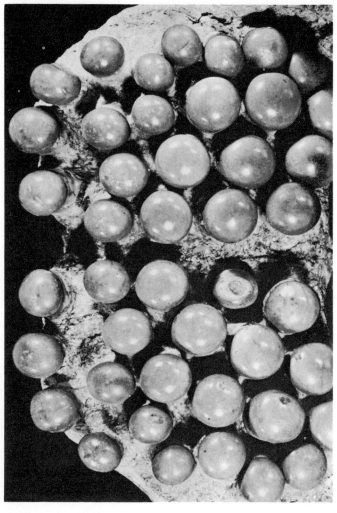

The lower jaw of a Mesozoic pebble-toothed fish from the Middle Jurassic.

The enormous brachiosaur from Tendagura in East Africa, exhibited in the Berlin Palaeontological Museum. Its size is suggested by the human skeleton in front of it.

Skull of a lizardlike form from the Cretaceous of North America.

Evolution is not Reversible

The fantastic notions about the great land saurians which had been cherished in the Cuvier era gave way in the era of Darwin to a sober, far more realistic view. The real disclosures could not come, however, until American palaeontologists had discovered the vast dinosaur cemeteries of the New World. But even before the excavation and researches of American 'dragon hunters', the most famous giant reptile of Europe, the iguanodon, was exposed in its true colours by a brilliant Belgian palaeontologist.

In early pictures the iguanodon was represented with a horn on its head. Sometimes the horn grew from its forehead, sometimes from its nose. There was a reason for these discrepancies. When Gideon Mantell, the discoverer of the iguanodon, began systematically digging for these saurians, he found certain bones in a quarry near Maidstone (in May 1834) which reminded him of the skull buds in cattle. From this he concluded that the iguanodon had been horned.

There had in fact been a number of horned varieties of saurians; but as it happened the iguanodon was not one of these. The apparent horn bud was in reality the animal's thumb, which ended in a cone-shaped spine. No one knew this, however, until 1877, when miners in a Belgian coal mine found the remains of seventeen well-preserved iguanodon skeletons, and the fragments of twelve more.

This was one of the most sensational saurian finds that had ever been made in Europe. The skeletons were buried more than a thousand feet underground, in a layer of marl from the Cretaceous; the miners had come across it while digging a new shaft. Fortunately they instantly reported their discovery to the manager, who happened to know something about fossils and fully realized the importance of the find. But he also realized that the fossils could be extracted only with the greatest difficulty. After a thorough study of the situation, and much consultation between the mine directors and the scientists of the Royal Museum of Natural History in Brussels, they decided to saw the entire layer of marl into blocks of one cubic metre, to encase each block in plaster, and to number them. Block after block was brought out of the mine, transported to the Brussels Museum, and there set up in the proper order. Thus the palaeontologists followed much the same procedure which was to be used nearly a century later by the archaeologists who cut the temple of Abu Simbel and other ancient Egyptian monuments into blocks in order to save them from the water backed up by the Aswan Dam and to reassemble them on another site.

Once in the museum, the plaster was removed from the blocks and the skeletons carefully chipped out. This task was directed by a young man who was only twenty years old at the time of the find. His name was Louis Dollo, and he was already the reptile specialist of the Brussels Museum. His colleagues at the museum were highly annoyed that such a 'green beginner' should be entrusted with so formidable a scientific undertaking. And when Dollo's first publications on the iguanodons of Bernissard stirred a sensation throughout the world of palaeontology, his envious colleagues succeeded in having him banished to the Museum cellar, where he henceforth had to work in a tiny, close room. The window of this cell-like place was barred; there was scarcely any heat; and the only illumination came from a feeble gas lamp. Dollo had to build his own tables and shelves out of crude boards.

In spite of these discouraging conditions, the young Belgian produced a masterly study of iguanodons and the other fossil reptiles of Belgium; the publication of this book alone made him at one blow the top palaeontologist in his country. In it he set forth his famous dictum: 'Evolution is not reversible.' According to this principle, known as Dollo's Law, characteristics which an animal has lost in the course of its family history never recur.

Exposing the Iguanodon

Dollo introduced the idea that species and whole orders of

226

animals can die out from over-specialization. When animals have adjusted so completely to a particular environment that they have lost the capacity to live in other conditions, they are no longer able to change back to an earlier state when the environment changes; they inevitably die out.

Dollo worked for almost twenty years cleaning, mounting, and describing the iguanodons of Bernissard. He was by then regarded throughout the world as one of the greatest experts on saurians. He determined that the iguanodon did not walk on all fours, but upright on two legs like the kangaroo and the predatory saurian *Megalosaurus*. It did not fight other saurians, but fed peacefully on conifer shoots, especially on araucaria. It used its long tongue to grasp the twigs, then cut them off with the beak-like prolongation of its mouth.

Dollo also considerably reduced the size of the iguanodon. This saurian had unusually large cheek teeth. Before Dollo's time the size of the body had been deduced from the size of the teeth; it was assigned a length of one hundred feet. When the first massive thighbones of the iguanodon came to light in England, the estimates were somewhat scaled down to between sixty and seventy feet. For a long time the iguanodon was regarded as the largest animal that had ever lived, or at least – as Friedrich August Quenstedt wrote in 1856 – the largest animal of ancient times: 'The mass of the clumsy body must have exceeded that of all prehistoric animals, even though it did not attain that of the whale.'

The excavations in America and Africa later revealed that creatures of this size had existed in the Upper Jurassic and Lower Cretaceous: colossal dinosaurs more than seventy-five feet in length, the largest four-footed vertebrates of all time. But the iguanodon was not in this class. From the tip of its tail to the end of its snout it was 'only' about thirty feet long. What the function of that curiously spiny thumb could have been, even Dollo was unable to explain. Perhaps it served as a kind of dagger when the mild reptile had to defend itself against predatory saurians; perhaps it also served as a kind of shears for cutting twigs.

Dollo's superior, the director of the Brussels Museum, was by no means delighted with the achievements and successes of his reptile specialist. One day he 'relieved' Dollo of his labours on the Bernissard finds and instead put him in charge of the museum's collection of fossil fish. But in this field too, Dollo, still working in his dreary cellar, immediately distinguished himself, making basic discoveries about the genealogy of lung-fish. His successor in the saurian department, on the other hand, could not handle the work on the iguanodons. Eventually Dollo was permitted to return to the saurians.

In 1904 he at last was able to leave the cellar for good, having been made director of the museum. From now on he devoted himself to the sundry relationships between organisms and their environments – that is, to developing Dollo's Law. The future belongs to the unspecialized; the highly specialized creatures sooner or later lose their ground – this was the thesis Dollo repeatedly upheld whenever the florescence and extinction of the saurians and other prehistoric animal families was discussed.

Flying Dragons over the Gingko Forest

It was one of Thomas Henry Huxley's inspirations to link birds genealogically with reptiles. What separates birds from reptiles, he noted as early as 1861, is almost exclusively the capacity for flight: the plumage, the transformation of the forelimbs into wings, the strengthening of the musculature of the chest, the formation of hollow bones and air sacs to diminish body weight. Otherwise, Huxley said, birds are little more than 'glorified reptiles'.

But what group of reptiles represent the ancestors of the birds? Huxley suggested the small reptiles which seemingly moved by running or hopping on their hind legs; various species of these *Pseudosuchia*, found in the Triassic strata of Scotland and South Africa, had beak-like mouths and long thin legs that made them look already partly like birds. But other palaeontologists thought it unnecessary to look among land reptiles for avian ancestors. After all, there was a group of saurians who had conquered the air as completely as the birds: the flying saurians. Why should the birds not be derived directly from them?

As it turned out, Huxley had been right; the flying saurians had nothing to do with the genealogy of birds. Although they stemmed from the same branch of the class *Reptilia*, they formed an independent line, and their method of flying was also entirely different from that of the birds. But it is understandable that for a while some palaeontologists wanted to connect them with the birds. For one of their principal sites is likewise the site where the most ancient bird has been found – the Jurassic slate in the vicinity of Eichstätt and Solnhofen.

In the Bay of Solnhofen

In 1738 the valuable limestone of Solnhofen in the Altmühl Valley of Franconia was discovered. Twenty-five years later

The reptile that resembled an ostrich, *Struthiomimus altus*. Many dinosaurs walked on their hind legs, had beak-like mouths, and in other respects resembled birds.

the inventor of lithography, Aloys Senefelder of Prague, found that Solnhofen slabs were particularly good for his new art of printing from stone. Subsequent exploitation of the lithographic stone brought so many fossils to light that Solnhofen soon became a meeting place for collectors and scholars from all over the world. The greatest palaeontologists devoted much time to the finds in the Altmühl Valley – among them Cuvier, Agassiz, and Owen. In the course of time they were able to say why such a vast number of different fossils were being found at Solnhofen.

Apparently, here too – as at Holzmaden – there had been a funnel-like bay almost completely surrounded by land during Upper Jurassic time. This bay must periodically have dried up. Along the shore grew araucaria and gingko trees, and vast stretches of reeds. The shallow waters teemed with seaweed.

For the followers of Darwin, the fauna of Solnhofen held special interest for the number of animals which had survived the ages in only slightly altered form, as 'living fossils'. Solnhofen proved that the ages and the fossils could not be delimited so precisely as had hitherto been thought. Among the characteristic inhabitants of Solnhofen Bay were the handsome sea lilies, which looked almost like flowers; the peculiar king crabs, which in reality are related to the arachnids rather than the crabs; and the Crossopterygia, those highly significant ancient fish which were closely related to the ancestors of the tetrapods.

Cuvier and his associates regarded all these animals as typical creatures of the Mesozoic era. But between 1836 and 1862 oceanographers such as Vaughan Thompson, Michael Sars, and Sir Charles Wyville Thomson, using novel dragnets, fished a number of species of sea lilies, very much alive, out of the depths of the oceans. The king crabs exist, as zoologists have determined, in two species along the coast of North America and in the vicinity of the Moluccas. And a surviving specimen of crossopterygian, *Latimeria chalumnae*, which strikingly resembles *Undina acutidens* from Solnhofen, has only recently been discovered in the seas off East London, South Africa.

Ichthyosaurs and plesiosaurs seem to have visited the bay of Solnhofen rarely, to judge by the sparse finds. But there are numerous species of primitive crocodiles and marine tortoises represented in the Solnhofen strata. As for land saurians, there are only small species, none longer than a cat. As the imprints of the slabs of stone show, some of these were caught by the tide on the beach and covered over by the ooze. A slab discovered by the German palaeontologist Hermann von Meyer in 1850 shows a homaeosaur struggling against just such a death. The imprint clearly shows how the animal tried to bend its body upward to escape suffocation. This homaeosaur was a close relative of the tuatara, the rhynchocephalous reptile still found on some of the islands of New Zealand.

Giant dragonflies and other insects swooped over the marshes, but the skies were largely dominated by flying saurians. The Altmühl Valley appears to have been a paradise for pterosaurs. Nowhere else in the world, aside from North America, have such numbers of pterosaurs been found.

The Phenomenon of Convergent Evolution

Hermann von Meyer, who found the suffocated homaeosaur, made a particularly intensive study of Solnhofen fauna. In addition to the pterodactyl, which Cuvier had seen, he described many new genera and species of pterosaurs. It is due to von Meyer that the fantastic notions of 'flying dragons' were gradually replaced by correct descriptions of the flying saurians' appearance and habits.

Hermann von Meyer was, however, still a scientist of the old school, an admirer of Cuvier, 'pervaded by the beauty of his views', as he put it. To the Cuvier law of correlation he added a supplement that was to play a great part in palaeontological research: 'I found that in one or several parts a creature could have correspondences bordering on similarity with another creature, without being related to it.'

This phenomenon of correspondence crops up in a very wide variety of animal groups: ichthyosaurs resemble dolphins; blindworms, snakes; swifts, swallows; opossums, rats. Thus it is merely a convergence, not a real kinship, that pterosaurs resemble birds in some respects and bats in others.

Hermann von Meyer might have become a German Cuvier but for an unfortunate physical disability. He had clubfeet and was therefore unable to stand or walk for very long. He was extremely sensitive on this score and refused to take any position in which the work could not be done sitting down. A passionate democrat, Meyer represented his native city of Frankfurt am Main in the German Bundestag. He served as comptroller and later as cashier-bookkeeper of the Bundestag – posts which took up most of his energies, to the detriment of his scientific labours. Even after he had become Germany's foremost vertebrate palaeontologist, author of a five-volume *Fauna of the Ancient World* which earned the highest praise in scientific circles the world over, he continued to check the income and outgo of the Bundestag, so that he had only his nights for examining and describing newly discovered fossils.

A sterling example of convergence: three unrelated animals (shark, ichthyosaur, and dolphin) resemble one another because of the similarities in their mode of life.

'A Truly Gigantic Dragon'

The pterosaurs of Solnhofen, which Meyer investigated and described, were small animals, the size of sparrows or at most ravens. They flew, Meyer determined, with the fluttering motion of the bat. In 1871, however, more pterosaurs were discovered in the Cretaceous strata near Cambridge, England; these too were small, but of different appearance. Unlike the Jura flying saurians, they had no teeth but long beaks instead. This fact led Harry Guvier Seeley, the English palaeontologist, to revive the theory that pterosaurs and birds might after all be closely related. Seeley, an enthusiastic follower of Lyell and among the foremost experts on reptiles in England, defended this view for

233

years against Huxley, although it soon developed that the small-beaked pterosaurs were in reality related to quite different, truly monstrous creatures that must indeed have looked like flying dragons.

In 1872 Othniel Charles Marsh – of whom we shall have more to tell – first came upon the remains of the strangest animal that had ever flown. A bone barely eight inches long, lying in Cretaceous strata of a dried river bed in western Kansas, put Marsh on the track of this flying monster. Attached to the bone was a curious joint of a kind Marsh had never seen before. It looked like the tibia of a huge bird; but no known bird possessed such a joint, for it moved in a manner that would have been quite useless to any well-constructed bird.

Marsh compared the bones with innumerable other fossils, and finally decided that the only joint like it he could find in a living or extinct animal was the joint on the finger bone of the pterodactyl. The pterodactyl was a quite small animal, whereas the American pterosaur must have been a giant, if the bone discovered by Marsh actually was a finger bone. Marsh was daring enough to predict, on the basis of this one find alone, that flying reptiles with a wingspan of over twenty feet would soon come to light.

That prediction was rapidly fulfilled. In the following years Marsh accumulated some six hundred remains of the skeletons of this huge animal – *Pteranodon*. It had a beak like that of the small pterosaur of the English Chalk; the forelimb measurements indicated that it had been a gliding flier. 'A truly gigantic dragon', Marsh called this largest winged creature that had ever lived on earth.

To this day we know comparatively little about the habits of the pterosaurs. It is assumed that they lived close to the seashore and preferred flying over water. The small European Jurassic and Cretaceous pterosaurs presumably captured all kinds of marine animals. The flying dragons of the American Chalk could probably hover over the open ocean like albatrosses and other great sea birds; they must have been excellent fishers of the high seas.

It created a sensation when the German palaeontologist Wanderer decided in 1908 that pterosaurs in all probability had possessed hairy pelts like mammals and had been warm-blooded.

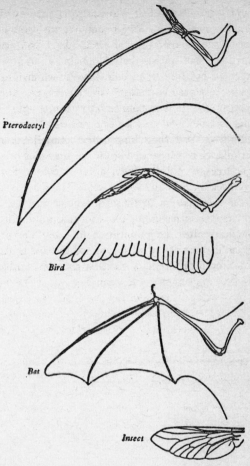

Four different types of flight: wing of a flying saurian, a bird, a bat, and an insect.

Twenty years later the pterosaur specialist Broili actually discovered the imprints of a delicate coating of hair on the neck, back, and wings of several European species. Whether the flying reptiles were warm-blooded remains a debated question. The American palaeontologist Edwin Colbert has commented that the pterosaurs must have been in active motion over considerable periods of time. For a cold-blooded animal that would be a difficult if not impossible feat, but not for a reptile that was as warm-blooded as a bird or a mammal. Warm-blooded animals can regulate the temperature of their blood by means of a complicated mechanism, so that it remains approximately the same under all conditions; in cold-blooded animals the blood assumes the temperature of the environment.

Many other questions in connexion with pterosaurs have remained unanswered. Did they lay eggs or bring their young into the world alive? Were they diurnal or nocturnal animals? Did they sleep like bats, suspended from the branches of araucaria and gingko trees, or did they mass in huge colonies on reefs, like present-day sea birds?

Whatever the answers to these questions may be, it is established that the ancestors of birds looked nothing like them. Only two years after the publication of Darwin's theory, the same Solnhofen quarries which contained the fossils of so many pterosaurs produced a relic of the first ancestral bird.

The Dispute over the First Bird

When Thomas Henry Huxley tried to trace the derivation of birds from reptiles, he drew a sketch of a hypothetical primordial bird. It was a feathered creature with reptilian teeth, claws on the ends of the wings, scales on its body, and a long, lizard-like tail. Huxley had no idea that almost immediately afterwards precisely such a primordial bird would be found, an animal that corresponded in almost all details to his drawing. The only clue that existed at the time he wrote was the imprint of a small bird's feather. With the discovery of that feather, however, a story began that at times seemed like a comic detective thriller.

The feather was found on a slab of stone from Solnhofen, and had been in the possession of Hermann von Meyer since 1860. Actually, the small imprint did not show very much. The interesting aspect was that it came from the Jurassic period, that is, that the animal had flourished in the Mesozoic. Hitherto palaeontologists had assumed that birds had not appeared on earth until the Tertiary period, at the beginning of the Palaeocene epoch. But the owner of this feather must have lived in the great days of the saurians. Consequently, the modest Solnhofen feather was bound to shake existing views on the age of birds and Cuvier's rigid periods of creation as well.

Meyer called the bird, on the basis of the single feather, *Archaeopteryx lithographica* – freely translated: 'Ancient feather from the lithographic stone.' There was nothing more that he could say about it. But he hoped that further remains of the same creature would soon be found.

A year later this hope was fulfilled. From a depth of sixty-five feet the workmen in the Solnhofen quarry brought up a slab which showed a nearly complete skeleton, minus the head, of the ancient bird. But Hermann von Meyer was out of luck. A fossil-collecting physician, Karl Friedrick Häberlein, quickly bought the slab; for he realized at once that he could turn it into a highly profitable piece of business.

The 'Griffin'

This Doctor Häberlein, who was the district health officer of Pappenheim, knew quite a bit about fossils. He also knew how eagerly scientists had been waiting for the appearance of a relatively complete archaeopteryx ever since the discovery of the Solnhofen feather. Moreover, he had heard of Darwin's theory and concluded that a 'connecting link', an animal showing the relationship between reptile and bird, ought to be worth a good deal of money. And the bird on the slab certainly looked like such a link.

Doctor Häberlein therefore decided to offer his ancient bird to various museums and scientific institutes – on terms, however, that had hitherto scarcely prevailed in the fossil trade. He invited the scientific agents of the great collections to visit briefly, so that they could inspect the find; then he waited for their offers, rejected each as it came, and in this way tried to drive up the price. What is more, although he permitted the scientists a brief glance at the precious slab, he barred them from taking notes, let alone drawing the bird.

For about a year Häberlein played this game with the palaeontologists. The result was that many naturalists denounced the Solnhofen bird as a fraud. Since Häberlein obstinately refused to permit serious scientific examination of the find, he was accused of forgery and deception Some of the animus rubbed off on Darwin's followers; they were said to have tried to prop up their sagging theory of evolution by manufacturing a monster.

Doctor Häberlein realized at last that he had gone too far and was preparing to accept the next offer that came along, when suddenly an excellent drawing of the archaeopteryx was published. All his mystifying had proved useless. Every palaeontologist now knew what the find looked like.

A clever palaeontologist with an excellent visual memory and a superior talent for drawing outwitted the cunning doctor. The man was Alfred Goppel, curator of the Munich palaeontological

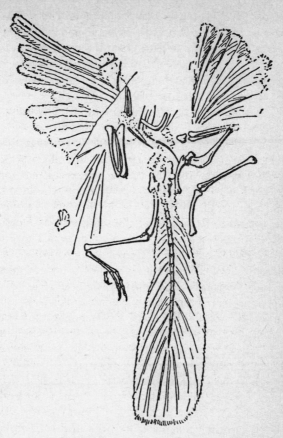

The London specimen of the *Archaeopteryx* – a fossil that became
world-famous. (Drawing from a plaster cast.)

collections. Goppel was an expert in Jurassic fossils; we are in-
debted to him for the customary Central European divisions
of the Jurassic period into the Lias, Dogger, and Malm series. At
the request of the Munich Museum he had visited Dr Häberlein
several times and impressed on his memory all the details of the
archaeopteryx slab. After each visit he scurried back to his hotel
and drew in all the details he remembered. Thus he gradually
compiled a picture of the primitive bird that was an astoundingly
good replica of the skeleton on the slab.

Goppel took this drawing to Professor of Zoology Andreas Wagner in Munich and asked him his opinion. What Wagner saw was a drawing of a creature about the size of a pigeon, with the head missing; the forelimbs were equipped with recognizable pinions. The exceedingly long tail was likewise feathered. The length of the tail startled Wagner; for although birds very often have long tail feathers, the tails themselves are only small stumps. The archaeopteryx was quite different; the tail was as long as that of a reptile, and from the root to the tip there grew from it, symmetrically on both sides, large vane-shaped feathers.

In other respects, too, the archaeopteryx differed from the typical bird. These differences led Wagner to conclude that it was not a bird at all, but a very odd reptile which strangely bore feathers instead of scales. Still, the creature was indubitably birdlike. To suggest what a problem it represented to palaeontologists, Wagner dubbed it *Griphosaurus problematicus*, 'the problematical griffin-lizard', and assigned it a place among its Jurassic contemporaries, the saurians.

In giving it this amusing name, Wagner had done the reptilian bird no good at all. For the scientists who doubted the genuineness of the find, or its significance, made fun of the 'feathered lizard'. Wagner's guardedness about an interpretation seemed an additional reason for extreme scepticism.

But while the Germans were wasting their time debating questions such as 'bird or lizard', and 'real or forgery', the English took rapid action. Richard Owen read Wagner's description and immediately sent one of his British Museum geologists, George R. Waterhouse, to Pappenheim. For £600 the archaeopteryx changed owners. Häberlein was glad to be rid of the troublesome slab, and had reason to be fully satisfied with the price. When representatives of German museums at last called on him once more, offering to resume negotiations, they learned that the bird or reptile had flown.

The disputes, however, now shifted to England. Owen and Seeley declared that in spite of the long reptilian tail the creature was a genuine bird and could not be construed as a piece of evidence for Darwin's theory. Huxley, on the other hand, felt that the discovery of this primitive bird was a personal triumph; to his mind the archaeopteryx was a perfect link between reptile and bird. And Darwin observed that the discovery of this bird with the long tail and claws on its wings once again proved how many gaps there were in the existing palaeontological record and suggested how many surprises there might yet be in store for scientists.

Darwinists and anti-Darwinists had a field day over the question of the missing head. If the archaeopteryx had had a toothless beak, as Owen and Seeley assumed, then it undoubtedly was a genuine although very primitive bird. If, on the other hand, it had had a reptile mouth with teeth, as Huxley and Dollo imagined, then it would have to be regarded as a transitional form between reptile and bird. In both England and Germany everyone waited in suspense for the discovery of another archaeopteryx.

Thirty-six Thousand Gold Marks for a Slab of Slate

Sixteen years after the Solnhofen find it came. A quarry owner named Dörr, who extracted slate from a hill near Eichstätt, fairly near Solnhofen, found an excellent fossil of an archaeopteryx – complete with head. Three fingers on each wing bore lizard-like claws. The head and thighs showed slight traces of scales. Otherwise, the animal was heavily feathered. The feathers on the wings, and the unusually long tail, were extraordinarily large.

Naturally, the head of the new archaeopteryx was of supreme interest to the scientists. It resembled the head of those small saurians, the pseudosuchians, which Huxley had called the ancestors of the birds, and also the dinosaur, *Ornithosuchus*, from the Triassic deposits of Scotland and the *Compsognathus* from the Jurassic strata of Solnhofen. Among these birdlike saurians the mouth was still lizard-like; in the archaeopteryx, however, it was already evolving towards a beak-like form. Both animals had the same pointed reptilian teeth in their jaws.

But before these details could be known, the same farce took place that had been played out after the discovery of the first ancient bird. Quarry owner Dörr did not sell the slab to a scientific institute, but to the same Häberlein family who had speculated on the first archaeopteryx. The old doctor had meanwhile passed away; but his son Ernst Häberlein had an even greater talent for commerce. He announced that the slab was for sale for the fearsome figure of thirty-six thousand gold marks. The price was tripled because this bird had a head.

No scientific institution in Germany was able to raise such a sum. The palaeontologists sent petitions to the government of the Reich, pointing out that unless immediate action were taken this precious specimen would probably be sent abroad. The Ministry of Education, which had charge of such matters, excused itself on the ground that the federal government had no museums; such a purchase was a matter for the individual States. But the State governments were not prepared for such an outlay; they pleaded that funds were lacking. In reality, none of the politicians in the cabinet of the Reich or in the State governments recognized the importance of the find.

Häberlein was finally prevailed on to give the Freie Deutsche Hochstift in Frankfurt am Main a six-month option to purchase the specimen. This institution, which was charged with the preservation of historically important antiquities (among other things it had bought the birthplace of Goethe in Frankfurt, to save it from demolition), then endeavoured to raise the money for the archaeopteryx. But it could find no generous donor; it had to let the option lapse. At this point a German naturalist who was professor of zoology at the University of Geneva intervened. He was Karl Vogt, one of the more passionate defenders of the theory of evolution. Vogt had taken part in the Revolution of 1848 as a member of the extreme left, and had subsequently emigrated to Switzerland. There he fought for everything that was new, interesting, and progressive: for Darwin, for oceanographic, Ice Age, and Polar research, for naturalism in literature, for Freethinking, and for a new social order.

The tolerant Swiss did not take these activities amiss. They even made Karl Vogt a member of the Federal Nationalrat. This restless spirit now got in touch with Häberlein, determined to obtain the precious fossil. He was understandably eager to have such a splendid demonstration of Darwin's theory in Geneva. And since he had great influence in his new country, he felt confident that he could raise the needed sum.

At a congress of the Swiss Naturalists' Society in St Gall, Vogt described the unique importance of the primitive bird and the benightedness of the men in charge of things in Germany. But his speech, which was intended to persuade the Swiss to give money for the purchase of the slab, soon wandered off – given Vogt's temperament, that was to be expected – into a violent political attack on Germany, Bismarck, Kaiser Wilhelm I, militarism, and imperialism. The case of the archaeopteryx was still another proof of the cultural barbarism of the German monarchy. Kaiser Wilhelm had plenty of money for soldiers and cannon, but not a mark for science.

Vogt was in for a bitter disappointment. The Swiss naturalists heard him out and then went home, without showing any stormy enthusiasm for the acquisition of the slab from Solnhofen. In democratic Switzerland, too, people had no money to spare for Herr Häberlein. Nevertheless, Vogt's speech had its effect – not in St Gall and Geneva, but in Berlin.

The industrialist Werner von Siemens, founder of electrotechnology in Germany, was stung into action. He haggled Häberlein down to twenty thousand marks, purchased the precious specimen, and thus secured it for his Fatherland. Shortly afterwards Siemens sold the archaeopteryx for the same sum to the Mineralogical Museum of Berlin University. There it rapidly became an object of national pride. The fossil was, as Germans were careful to point out to the British, not only much more handsome than the first Solnhofen find; it was also an entirely different and far more interesting type than the piece in London. Wilhelm Barnim Dames, palaeontologist at the Berlin Museum, paid tribute to Siemens in baptizing the new primitive bird with his name. Later, in fact, it was classified as a special genus: *Archaeornis siemensi*. More recent studies, however, have proved conclusively that the two birds belonged to the same species.

The question whether this evidence for Darwin's theory was to be called a genuine bird or a transitional form between reptile and bird continues to be disputed to this day. The most apt description of the creature was coined by Gerhard Heilmann, the Danish zoologist, who called it 'a warm-blooded reptile disguised as a bird'.

Seventy-nine years passed after the Eichstätt find before another archaeopteryx came to light. It was found in 1956, only two hundred yards from the spot where the first specimen had been discovered. This archaeopteryx also lacked a head; but the limbs and feathers were especially well preserved. F. Heller, the German palaeontologist who examined the specimen, found that its vertebrae and hollow bones are typical avian bones. The old question of whether the creature could fly properly or only flutter was also cleared up at this time. Earlier theories were that the bird more or less parachuted from tree to tree, only occasionally moving its wings clumsily to achieve fluttering flight. But Heller perceived by the characteristics of the wings and the plumage that the animal had really been capable of some flight, albeit awkward.

In all other characteristics, however, the new find proved itself an ideal example of those intermediate forms which Huxley had postulated and which led from Mesozoic reptiles to present-day birds.

Mass Migrations and Mass Graves

Shortly after the publication of the theory of evolution, it developed that the fine, seductive order of geological ages and their fossils, which scientists had been firmly subscribing to, was little more than a pretty fiction. Many animal forms and groups appeared suddenly in periods in which, according to existing theories, they had no business to be.

There had already been birds in Jurassic time, primitive mammals in Triassic time. Some giant mammals of the American Tertiary lived on into the Ice Age and the postglacial period. Mesozoic and Tertiary animal species exist to this day, often in almost unchanged form – the so-called 'living fossils'.

The whole science was in flux. Fossils could no longer be dubbed as belonging to particular ages; they had to be studied anew, in the light of earth's history. Evolutionists began sketching family trees to show how various animal forms had developed stage by stage from simple infusorians up to man. Fossil animals had to be placed in these family trees.

Some fossil creatures actually proved to fit perfectly into such hypothetical stepladders of life: the Silurian ostracoderms (shell-skins) as ancestors of the fish; the Devonian crossopterygians as ancestors of the amphibia; the Carboniferous armoured amphibians as ancestors of the saurians, the Permian-Triassic theriodonts, a primitive group of saurians, as ancestors of the mammals; the archaeopteryx, already discussed; and finally Neanderthal man and *Pithecanthropus erectus*, the 'ape-man who walked erect' whom Ernst Haeckel had predicted and who was found in 1891 by the Dutch military physician Eugen Dubois, on Java, exactly where Haeckel had surmised.

But such transitional forms did not at first appear with anything like the frequency which Darwin's followers had hoped for. That was why many scientists considered it perilous to postulate long pedigrees on the basis of the few fossil connecting links that were known. When Ernst Haeckel adorned his book *Natural History of Creation* with a multitude of such pedigrees, the Berlin physiologist Emil Dubois-Reymond declared that these animal family-trees were 'worth no more than the genealogies of Homeric heroes'.

Dubois-Reymond had good reason for this remark. Most of the family trees consisted for the most part of gaps, at least in the fossil part of the line. Here and there a solitary fossil would be entered. No wonder that well-meaning naturalists, while they might have nothing against the theory of evolution in itself, smiled at these constructs and observed that they contained a great deal of poetry and very little truth.

Many of the fossil sites seemed to confirm Cuvier in his theory of totally destructive geological revolutions. The limestone quarries at Holzmaden and Solnhofen, the mammoth graveyards in Europe and Siberia, the enormous collections of bones in Patagonia, were not the only mass graves of fossils to be found. In Greece, in northern India, in the Mongolian desert, in Africa, and above all in North America enormous storehouses of bones were discovered, like vast cemeteries for antediluvian animals.

The Treasure Hunters of Pikermi

A particularly interesting mass grave of Late Tertiary animals, which was to prove highly illuminating to Darwin's followers, has been found in the classical countryside of Greece, near the small village of Pikermi, fairly close to Athens. Here, beneath the strewn rock of Megalorhevma Brook, the British archaeologist George Finlay, hunting for Hellenic antiquities, in 1835 found a few fossil bones. Three years later Pikermi suddenly became a focal point of interest as the result of a comic chain of circumstances.

The King of Greece at that time was Otto von Wittelsbach. Among his troops was a Bavarian soldier who happened to find a skull and some weathered bones in the valley of the Megalorhevma. The skull looked fairly human; moreover, there were pretty calcite crystals sparkling in the crevices of the bones. The

Bavarian decided that he had found diamonds, and took the finds home with him on his next furlough. Back in Bavaria he boasted so loudly in taverns about his luck that the police became interested. One day a rough police sergeant laid a hand on his shoulder and arrested him on the charge of having committed grave-robbery in Greece.

Fortunately for the soldier, the police asked a scientist to examine the skull and bones. For the Munich police apparently were not entirely convinced that the skull was actually human and that the crystals were diamonds. The scientist was Andreas Wagner, who was later to be involved in the archaeopteryx affair. Wagner's verdict exonerated the soldier, for he found to his astonishment that the bones belonged to an ape of the Tertiary period.

That was, in 1838, a real surprise for science. For hitherto palaeontologists had bowed to Cuvier's dictum that apes as well as man could not have appeared until the present period of geologic history. The *Mesopithecus pentelicus*, as the Pikermi ape was named, exposed this error. It was only a cynopithecus (black ape), which is actually a monkey; but it set off the search for higher apes considerably closer to man.

About twenty years passed, however, before a kind of general inventory – prompted by Darwin's theory – was taken of all the ape-like fossils so far collected. A number of finds had been regarded as doubtful. These were now examined more closely, and it turned out that not only monkeys but anthropoid apes also had lived in the Tertiary period. It was not until the twentieth century, however, that palaeontologists achieved real clarification of the genealogical relationships between fossil anthropoid apes and prehuman forms.

A Terrible Wild Beast

After the identification of the bones, Pikermi became a mecca for diggers and fossil collectors. Greek, French, German, Austrian, Swiss, and British scientists made the pleasant pilgrimage to Attica year after year. From the rubble of the brook they dug up animals which seemed partly very strange, partly remarkably like the large fauna of the African plains today.

Among the long-ago animals of Pikermi which are now entirely extinct were the chalicotheres. They were large rooting animals, distantly related to the ungulates, whose mighty claws probably served to dig up tubers and bulbs. The word means 'terrible wild beast'; and the name is an apt one in both the literal and the figurative sense. For up to that time this strange Tertiary monster had been a terror to all palaeontologists.

Skulls and bones of the dinothere had been found frequently since 1829, especially in the Middle Tertiary sands of Eppelsheim near Mainz. Scientists wondered especially at the huge tusks, which were not enlarged upper teeth, as in elephants, but lower teeth. Moreover, they arched down and backwards. Standing more than fifteen feet high, the dinothere was the largest land animal known to man; it was decades before the primitive rhinoceros from Central Asia, the baluchithere, was found to be larger.

The discoverer of the Eppelsheim bones, the German palaeontologist Johann Jakob Kaup of Darmstadt, classified the huge beast as a kind of hippopotamus. The last proponent of the Deluge theory, William Buckland, considered it a relative of the walrus. Other scientists compared it with the mastodon and megathere; they suggested, with patent uncertainty, that the mysterious creature might be a transitional form between the elephant and the giant sloth. Even Cuvier was at a loss. He classed the dinothere first with the tapirs and then with the sea cows.

The dinothere offers a significant example of the sharp differences of opinion that often prevailed among leading specialists before the age of Darwin. When more remains of the animal were found in Pikermi, and soon afterwards in Rumania, India, East Africa, and other regions, it turned out that the dinothere had not been a strange monster living in water like a walrus or sea cow, but a peculiarly specialized relative of the elephants adapted to conditions in swampy forests. The odd tusks in the lower jaw were possibly used as a kind of hoe, to clear a path through dense underbrush.

When the Middle Tertiary swamp-forests vanished, these large proboscidians were no longer able to adapt to the new conditions. They became victims of Dollo's principle, that evolution is not reversible. In the recent Tertiary period, which corresponds to the age of the Pikermi finds, they died out as a result of their over-specialization.

Actually, they did not quite belong among the Pikermi fauna. For Greece at that time was a plain covered with low growth, like present-day East Africa. The dinotheres, like other swamp-loving trunked animals of Pikermi, could probably find a sparse living only in a few oases and river valleys. The increased drought steadily narrowed the area in which they could survive. On the other hand, the denizens of the parched steppe multiplied. Pikermi in Greece became very much of an 'African' animal paradise.

Like the African Plains

The foremost interpreter of the Pikermi finds, Albert Gaudry, has traced innumerable parallels between Greece in the Early Tertiary and Africa today. Other prominent palaeontologists have also envisioned Pikermi as a Tertiary Africa – among them Edouard Lartet, Wilhelm Barnim Dames, Melchior Neumayr, the Greeks Mitsopoulos and Skouphos, as well as the sponsor of the notorious Piltdown man, Arthur Smith-Woodward, and even a Prince of Orléans who happened to be interested in palaeontology. Their descriptions of the Pikermi region as it must have been would strike a familiar chord in a man who has been on a photo safari in the game preserves of Kenya or Tanzania.

Ancestors of the elephant grazed in Attica in the Late Tertiary, together with rhinos, ancient giraffes, zebra-like *Hipparion* horses, and an astonishing variety of antelopes and gazelles of all sizes and shapes. Ground hogs dug burrows in the soil of the plain; shrew-mice formed populous colonies in rocky areas; black apes swung through the trees of the savannas. And as in today's Africa lions and leopards lie in wait for their prey at the water holes, so the great sabre-toothed tigers lurked in the underbrush near water. Predators like hyenas, dogs, and civet cats cleaned up the carrion, just as they do today in East Africa.

Why should the present-day fauna of Africa so closely resemble the ancient fauna of Pikermi? Before Darwin's time there was no explanation for such a phenomenon. But with general acceptance of the idea of evolution, it became apparent that the animals must stand in close genealogical relationship to one another. Later a similar fauna, though somewhat older and on an even grander scale, was found in Asia, especially in the Sivalik Mountains of northern India. There, too, on the southern slopes of the Himalayas, vast herds of 'African' animals had once lived, even more variegated in species and forms than those of the African plains today. Tertiary Africa, on the other hand, was without most of the animals which are now typical of the Black Continent.

From these facts palaeontologists drew a conclusion which at first seemed distinctly bold, but which was later confirmed by many finds in Asia and southern Europe. Possibly the present fauna of the African plains is not of African but Asian and southern European origin. Today we know that many of the characteristic animals of Africa, especially the giraffes and antelopes, originally had their homes in Asia and migrated gradually to Africa by way of Greece millions of years ago.

In Pikermi the palaeontologists were particularly struck by a giraffe-like ungulate with a short neck and frontal bumps on its skull. Gaudry and Lartet named it after its discoverer Duvernoy *Helladotherium duvernoyi*, 'Duvernoy's Greek animal'. The evolutionary scientists were delighted that the fossil remains of their hypothetical giraffe ancestor had been found at last. In France, Lamarck's theory was at this very time being revived by Alfred Giard, the biologist and oceanographer. Lamarck had been particularly fond of citing the giraffe, who had come by his long neck through practice and adaptation. To those French evolutionists who for a time preferred Lamarck's ideas to Darwin's, the 'Greek animal' was a welcome fossil argument in favour of the great, unfortunate, and far too long forgotten founder of the milieu theory.

The *Helladotherium* was destined to create an even greater stir. In 1901 Sir Harry Johnston, the British governor of Uganda, discovered in the Congo forest a mysterious new type of large animal, a creature which seemed to fall somewhere between a zebra, an antelope, and a giraffe. British zoologists then found that this living marvel from darkest Africa, the okapi, was

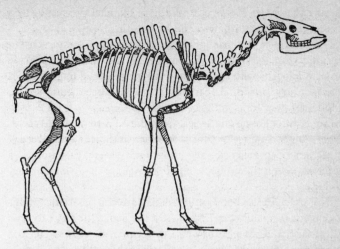

The reconstructed skeleton of the short-necked giraffe, *Hellado-therium duvernoyi*, of Pikermi: an okapi of Late Tertiary times in Greece.

virtually identical with the fossil 'Greek animal' of Pikermi. The okapi, too, is a 'living fossil' – a surviving cousin of the short-necked giraffe of Late Tertiary times, which has scarcely changed in the course of millions of years.

How had these vast quantities of animal bones accumulated in Pikermi? Some scientists assumed that heavy rains had caused the rivers of the plain to flood, and that the high water had floated the bones of innumerable animals that had died in the course of time into the gorge of the present Megalorhevma. Melchior Neumayr, the Austrian palaeontologist, recalled that Darwin had commented on the great droughts and subsequent high animal mortality in South America; something of the same sort might have happened in Tertiary Greece, he suggested. Other experts held that the valley of the Megalorhevma must have been an ancient eating place for predators. Here, they argued, the sabre-toothed tigers, hyenas, and other carnivores had carried their prey for many generations. Othenio Abel reasoned that the herds of animals might have been killed by periodic prairie fires such as still occur in Africa. Many animals, fleeing the flames in panic, would have plunged down the steep banks of the Megalorhevma and broken their legs.

The origin of this mass grave remains an unsettled mystery to this day. But the finds at Pikermi did for the first time provide palaeontologists with some idea of the great animal migrations of the Tertiary period. Apparently there had been innumerable such migratory trails in the course of the Tertiary – from Asia to Europe, from Europe to Africa, from the New to the Old World, from South America to North America and vice versa. Reconstruction of prehistoric migrations gave the palaeontologists information on the probable point of origin of many mammalian groups.

Not only giraffes but deer, it seemed, are also of Asiatic origin. Elephants originally came from Africa, whence they spread over all of Asia and into Europe and the Americas. Camels, on the other hand, originated in America, where they still exist as small, humpless species, the llamas and their relatives. Anthropoid apes probably arose in Africa. Perhaps prehuman forms, which likewise migrated into vast areas from southern Africa to southeast Asia at a very early period, were also of African origin; the question is still much debated.

Palaeontologists experienced a particular surprise when they began tracing the origin and migration of the horse. It was in connexion with this animal that the Darwinists enjoyed their greatest triumph.

The Darwinists' Darling

The incompleteness of the palaeontological record caused many evolutionists concerned with family trees to rely more upon the comparative anatomy, morphology, and embryology of living species than on the sparse remains of the ancestors of those species.

'The documentary value of palaeontology for the history of species is much overestimated,' Ernst Haeckel admitted sadly in one of his lectures. '. . . Fossils alone could never teach us anything about the great majority of animal and plant species that have inhabited our globe.'

It may have seemed an inconsistency, therefore, when in 1872 Haeckel suggested to one of his students, a thirty-year-old Russian named Vladimir Onufrievich Kovalevsky, that he write his doctoral dissertation on fossil horses. Haeckel could not suspect that during the next four years the fossils of horses would provide one of the most fascinating and complete palaeontological documentations of the theory of descent that anyone could desire. Rather, Haeckel merely hoped for a little more light on the ancestry of the horse, which at that time was but little understood.

Soon after the publication of *The Origin of Species* Huxley had sketched out a hypothetical pedigree for horses. He held that the horses derived from three-toed perissodactyles which must have been rather similar to the tapirs, and that in turn these three-toed proto-horses had evolved from four- to five-toed and probably very small animals. Thus the hoof must have developed quite gradually by progressive reduction of the toes.

This sounded plausible, and there was some fossil evidence for it. In a Lower Eocene stratum of clay near London Richard Owen had discovered a four-toed primitive ungulate scarcely the size of a fox, the *Hyracotherium leporinum*. According to Huxley this 'fox pony' might very well have been a very early ancestor of the horse. In the Early and Middle Tertiary of Europe three-toed tapir-like perissodactyles had been found; according to Huxley these represented a further stage of evolution. They were the *Palaeotherium magnum*, which Cuvier had extracted from the gypsum of Montmartre, and the *Anchitherium aurelianese* from the Miocene of Orléans, described by Hermann von Meyer. From the next period, the Pliocene (Late Tertiary), came the zebra-like *Hipparion mediterraneum* that had been excavated at Pikermi; it still possessed three toes on each foot, but two of them had atrophied to mere appendages. In the Pleistocene, at the beginning of the Ice Age, the first single-hoofed horses appeared alongside the hipparion.

So far, the whole picture looked fairly satisfactory. The European finds yielded a family tree which had, to be sure, sizeable gaps, but at least revealed a usable evolutionary line: from four-toed 'fox ponies' to three-toed 'tapir horses' and to the almost one-toed and finally genuinely hoofed horses. But unfortunately, across the ocean in North America, an altogether different picture was in the making.

From Browser to Grazer

From 1849 on, Joseph Leidy, a professor of anatomy in Philadelphia, had been publishing reports on a number of prehistoric horses found in the Tertiary strata of North America. The finds were almost all chance discoveries by farmers or prospectors which some thoughtful schoolmaster, minister, or sheriff had sent to Philadelphia. They excited great interest because they belied the old legend that horses had been unknown in the Americas until they were introduced by European settlers.

But there was another reason for the excitement. It quickly became evident that America in Tertiary time had also had four-toed, three-toed, and one-toed horses. Only the crude 'tapir horses' so typical of Europe had apparently never existed there. All American early horses, whether large or small, and from whatever period, seemed much more horselike than the corresponding Old World species.

The idea therefore arose that perhaps present-day horses evolved twice: in the Old World by the roundabout route of tapir-like forms, and in the New World by a more direct route of steadily increasing 'horsiness'. This 'theory of parallel evolution' was advanced with a great deal of acuity, although a follower of Darwin declared quite rightly: 'There is something altogether weird in the idea that the same animal should have arisen on two different continents through a series of different intermediate forms.'

Vladimir Kovalevsky – to return to him – was therefore doing his doctoral dissertation on a truly sensational topic, which might have seduced a less objective scientist into the byways of a metaphysical plan of creation. But Kovalevsky himself was a man quite as interesting as his subject. He came from a family of impoverished landowners and had been working as a translator since the age of sixteen, in order to raise money to attend school. Thus he was a working student – something virtually unknown in Europe in those days. He had to struggle not only because of his own economic situation; he also had taken a wife, though they had married only pro forma, in order to enable her to study abroad. She was Sophie Korvin-Krutkovsky, the daughter of a Russian general, who later developed into an eminent mathematician. The platonic alliance lasted five years; then, after the two had studied and starved together at a wide variety of universities in Western Europe, their connexion became a real marriage.

Kovalevsky expanded his doctoral thesis into a five-volume monograph of fossil ungulates – the first major work of this sort based on the theory of evolution. To assure his book the widest readership, Kovalevsky wrote it simultaneously in French, German, and English. Meanwhile he continued studying in Munich, Würzburg, and Berlin, travelled to the sites of many palaeontological finds, and visited Darwin in Down. In spite of all these activities, he took only two years to complete his vast project.

In the course of his studies Kovalevsky examined not only the number of toes but also the form of the teeth in order to determine the interrelationships of various equine species. In the Early Tertiary, he had learned from palaeobotanists, there were as yet no extensive meadows; hence the early horses must have been browsers, eating leaves. Not until the Middle Tertiary were there extensive grassy plains; and from this time on the early horses became grass eaters. Horses with simple, low-crowned teeth, indicating a browsing habit, must therefore have been

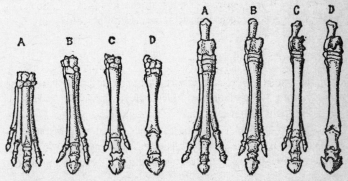

The reduction of toes in the horse. A = Eohippus (Eocene), B = Miohippus (Oligocene), C = Merychippus (Miocene), D = horse today.

lower in the family tree than those with complicated, high-crowned molars which could chew up hard grasses.

The conclusion that Kovalevsky drew from all his evidence has an inspired simplicity comparable only with Cuvier's deductions from analogies: certain characteristics of the primitive horses must necessarily go together and the animals could therefore have appeared only in particular epochs of the Tertiary period. The succession he worked out was as follows:

Eocene: Body the size of a fox, four toes, simple dentition, browser.

Oligocene: Size of a sheep, three-toed, simple dentition, browser.

Miocene: Size of a donkey, three-toed, two toes somewhat reduced, complicated dentition – transition from browser to grazer.

Pliocene: Size of a zebra, single hoof with appendages, specialized dentition, grazer.

Pleistocene: Size of a horse, single hoof, specialized dentition, grazer.

This new equine family tree displaced the clumsy 'tapir horses' of Europe, the palaeotherium and anchitherium, from their previous role as favourites of the palaeontologists. They could not have been direct ancestors of the solidungulate (one-hoofed) horses. They were too large and too unlike horses for that. Moreover, the anchitheria had still been browsers towards the end of the Miocene. As far as the ancestry of the horse was concerned, North America seemed the place to look.

A Complete Pedigree

For the finds in North America now began following hard upon one another. Othniel Charles Marsh, their discoverer – who also dug the giant winged saurians out of the Cretaceous strata of Kansas – was soon to become the uncrowned king of palaeontologists in America. We shall have much to say about him later on in connexion with the great age of dinosaur excavations in the West.

As early as 1868 Marsh had found the fossil skeleton of a three-toed horse at the site of a well being dug in the soil of a

dried-up Nebraskan lake. The animal was about as big as a police
dog, and its discovery led him to undertake four expeditions
with scientists of Yale University in order to search for more
American dawn horses. From 1872 to 1874 he and his associates
dug in Nebraska, Wyoming, and the Dakotas. They found some
thirty types of fossil horses, from all epochs of the Tertiary
period. *In toto* these formed so complete and well linked a chain
of evolution that Huxley remarked: 'It borders on witchcraft.'

The series began with the cat-sized *Eohippus* of the early
Eocene, more than fifty million years ago. This lilliputian horse
was a leaf-eater and had four toes on the forefeet and three toes
on the hind feet. It was followed in the course of the Eocene by
Orohippus and *Epihippus*, about the size of a fox terrier, which
also browsed on leaves. About thirty-seven to twenty-six million
years ago, in the Oligocene, appeared sheep-sized, three-toed
forms, *Mesohippus* and *Miohippus*; in *Meso-* and *Miohippus* two
of the toes were smaller than the central one. In the Miocene,
about twenty-six to seven million years ago, *Parahippus* and
Merychippus gradually shifted to life on the plains, and to grass
for provender, as their hoofs and molars indicate. In the Pliocene, from twelve to two million years ago, there appeared

Dawn man riding dawn horse. After a conversation with Marsh
in 1876 Huxley decorated his family tree of the horse with this
caricature. The drawing alongside shows how the toes developed
into the hoof, and the browser's teeth into grazer's teeth.

Hipparion, with the side toes reduced to appendages – the same species that also inhabited Europe. *Pliohippus*, which shared the plains with it, had complete but slender side toes. Finally, in American Pleistocene strata Marsh found the most recent of all: genuine horses of the genus *Equus*, to which present-day horses, donkeys, and zebras belong.

The history of the horse, which Kovalevsky had attempted to draw up by pure deduction, could now be traced from stratum to stratum, from form to form, thanks to Marsh's finds. Haeckel exulted, saying that the American fossil horses were the 'show horses of the theory of evolution'. Additional finds in North America closed many of the remaining gaps in the ancestral series. In some cases the transitions were so close that it was extremely difficult to separate species and genera.

'So complete a family tree,' Abel has commented, 'should be the *reductio ad absurdum* of systems seeking always for characteristic differentiating features. For we are almost obliged to place the dividing line of two species between father and son. The scientific names that have been proposed for the various stages signify only the individual steps in an evolutionary progression that leads upwards without a lacuna.'

When Marsh appeared in England and called on Huxley with boxes full of fossil horse bones, he and his host came to the conclusion that the race of horses had evolved only *once* – in North America. They also agreed that the European *Palaeotherium* and *Anchitherium* could not have been ancestors of present horses, but were instead curiously specialized descendants of emigrants from North America. About five times in the course of the Tertiary early horses migrated from North America to the Old World. Four times they took evolutionary dead ends and died without issue. The fifth time, in the Pleistocene, the horses at last established themselves in the Old World, but became extinct in the Americas.

Twentieth-century excavations have shown that the pedigree of the horse is after all somewhat more complicated than Huxley, Kovalevsky, Marsh, and Abel assumed. For there were a number of other dead ends. But the phylogeny of fossil horses, which Kovalevsky deduced by pure reason and Marsh found in actual bones, has remained to this day the ideal demonstration for the theory of evolution. Huxley once commented that Marsh had conjured up the very evidence that he, Huxley, had so intensely wished for.

Death of a Genius

Vladimir Kovalevsky, the deviser of the horse's family tree, had

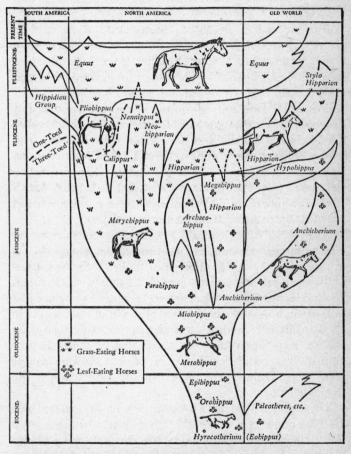

The descent of the horse and the distribution of various evolutionary forms over the Old and the New World. (After Simpson and Heberer.)

far less good fortune and success than the discoverer of fossil horses, Othniel Charles Marsh. Soon after writing his monograph on ungulates, Kovalevsky returned to Russia with his wife, determined at last to improve their economic position. But they became entangled in all sorts of speculations which turned out badly. Moreover, good society in Russia was at that time fiercely anti-Darwinist. The Kovalevskys were regarded as foreign-tainted freethinkers, and suffered much unpleasantness and mistrust.

Their struggle for livelihood became such a strain on their nerves that tensions arose between the couple, who had hitherto bravely sustained each other. Vladimir Kovalevsky more and more developed into one of those tragically divided souls such as Dostoevsky might have portrayed. Although he had a teaching position at the University of Moscow, he became a partner in a shady company engaged in dubious petroleum projects. Periodically, he would fall into depressive states; then again he would have self-destructive frenzies in which he tried to catch up on all that life had hitherto withheld from him. His wife left him and moved to Berlin. His interest in science dwindled to such a point that he scarcely paid attention to the great discoveries in America, which so brilliantly confirmed his ideas.

When the oil company finally went bankrupt, Kovalevsky put a bullet through his head. He was then forty-one years old. As it turned out, the suicide of this man, undoubtedly the greatest of Russian palaeontologists, proved a salutary shock to his family. His wife Sophie began to work seriously, and, with a post in Sweden, became the first woman professor of mathematics in Europe. His elder brother, Alexander, developed into one of the leading evolutionary biologists; among other things he discovered that certain curious, very primitive-looking marine animals, the tunicates and the amphioxus, have affinities with the vertebrates.

The genius of Vladimir Kovalevsky was fully recognized only after his death. Henry Fairfield Osborn, one of the greatest of American palaeontologists, commented that whenever a student asked him how to study palaeontology, he could do no better than point to Kovalevsky as an example.

Book Five

Dragon Hunters

Round about his teeth is terror. His back is made of rows of shields, shut up closely as with a seal. . . . His eyes are like the eyelids of the dawn. . . . In his neck abides strength, and terror dances before him. The folds of his flesh cleave together, firmly cast upon him and immovable. . . . Upon earth there is not his like, a creature without fear.

JOB, 41.14

Pioneers in the Wild West

Nowadays the great palaeontological vertebrate collections are in the United States. From the fortunes of Rockefeller, Carnegie, and other multimillionaires have flowed the funds needed for training fossil hunters, for equipping expeditions, and for conducting large-scale excavations. American palaeontologists like Marsh, Cope, Osborn, and Andrews have been so successful that they are regarded in Europe with as much astonishment as fabulous beasts.

Nowhere else in the world has the general public been so aroused to an interest in the past history and bygone creatures of the earth. Majestic museums of natural history have been built merely to house the bones of fossil mammals, toothed birds, and giant saurians which came to light. The museums were not only built but they are enthusiastically visited by hosts of ordinary folk, who are thus provided with considerable scientific education. Palaeontology, as the American biologist George Gaylord Simpson remarked, has in this respect made 'its greatest contribution to the cultural history of America'.

But things were not always thus. On the contrary, at the beginning of the nineteenth century Americans were by no means receptive to natural science. While the picture of the past as drawn by such men as Buffon and Cuvier was attracting interest in Europe, in America only a few eccentrics, curiosity collectors, or showmen were concerning themselves with old and often gigantic bones, which turned up with remarkable frequency in gravel pits, quarries, dry river beds, or other localities. Farmers and gold prospectors gave a utilitarian turn to these strange objects by making thresholds, clothes racks, and mantels out of them.

Great Bones on the Banks of the Ohio

The first serious fossil collector in America was George Croghan. A Dubliner, he had emigrated to Pennsylvania in 1741. By chance Croghan learned of the large petrified bones and teeth that had been found in Kentucky and offered to purchase any items of this sort. He soon had a large collection and wanted to learn more about his acquisitions. But he could find no one in America who could enlighten him. Croghan therefore sent his store of curiosities to London, where they were examined by various British experts. The bones and teeth were recognized as coming from elephant-like animals. Benjamin Franklin happened to be in England at this time. He compared Croghan's finds with the tusks and molars of Old World elephants and Siberian mammoths, and came to the conclusion that the 'Ohio mammoth', as the American type was called, constituted a distinct species, different in a number of points from other elephants.

Today we know that Croghan's bones were those of mastodons. Since Franklin suggested in his notes that the 'Ohio mammoths' need not necessarily be extinct, the rumour circulated that there were still American elephants in unexplored regions of the New World. Though this proved not to be so, the American mastodon, as old cultural strata and Indian representations testify, appears to have lived up to a few thousand years ago.

When word of the American finds reached Paris, French naturalists went to some lengths to encourage the Americans to continue collecting fossils, and to enter into exchange arrangements. In 1786 Lamarck and Geoffroy de Saint-Hilaire wrote a joint letter to the Philadelphia Academy referring to the 'giant bones on the banks of the Ohio' and recommending that attention be paid to such fossils 'since exact knowledge of them is more important for the study of the earth's history than is generally supposed'.

Eleven years later Thomas Jefferson took up this suggestion. He turned to William Clark, who later was to join Jefferson's private secretary, Meriwether Lewis, in the most celebrated expedition of the age. Jefferson, then Vice-President, suggested that

Clark set out to look for more 'Ohio mammoths' and other fossils. The region Clark combed proved to be the first of the many extremely productive fossil sites which have made North America so interesting to palaeontologists: the Big Bone Lick Region. Here Clark found, along with other specimens, some three hundred mastodon bones.

In order to awaken an interest in natural science, and especially in fossils, among his countrymen, Jefferson had Clark's finds exhibited in the White House, together with the bones of a giant sloth, and petrified fish, mussels, and other palaeontological specimens. He summoned Dr Caspar Wistar, one of the few experts in the country at this time, from Philadelphia, requesting that he make a thorough study of the exhibits according to the methods current in Europe.

This was a highly promising beginning for the development of palaeontology in the young American republic. But for the time being, the seed sowed by Jefferson did not sprout. Philadelphia alone became a centre of natural science. There Dr Richard Harlan gave scientific names to American fossils, arranging them on Linnaean principles and trying to interpret them by the methods of Cuvier. His writings contain the first word on American saurians, although he did not know how to identify them correctly.

But Dr Harlan remained a noble exception. Very few Americans could handle a geologist's hammer, and most amateur collectors knew nothing whatsoever about Linnaeus, Cuvier, or the methods of science. One such amateur, an earnest Quaker named Timothy Conrad, tried to replace the generally accepted nomenclature and classification by a poetic system which evidently seemed to him more graceful and less dry. He regularly invited poets and writers to his salon, showed them petrified sea shells he had collected, and invited them to compete with one another in devising pretty new names for his finds. Of course Friend Timothy held firmly to the story of Noah and the Flood.

'Give Trilobites a Tongue'

The indispensable basis for all palaeontological research is a thorough knowledge of stones and geological strata. For many years America lacked that basis. In 1806 the first notable American geologist, Benjamin Silliman, took charge of the department of chemistry and mineralogy at Yale University, then still a modest college. At that time the concept of 'geology' was virtually unknown, as Silliman later wrote in his memoirs. The entire mineralogical collection of the college consisted of a bushel of unmarked stones.

It was not until the middle of the nineteenth century that Silliman's son-in-law, James Dwight Dana, raised geology to the rank of a science in America. Dana had followed developments in Europe; he took part in a number of major expeditions, and like Darwin went on a four-year voyage around the world under the command of naval officer Charles Wilkes. Again like Darwin, he amassed a great deal of knowledge and experience during this voyage. But he had a far harder time of it than Darwin under Captain Fitzroy. Lieutenant Wilkes seems to have regarded the scientific enthusiasm of Dana as foolish, and the professor himself as a thorough nuisance. When Wilkes decided to attempt circumnavigation of the Antarctic continent, it did not occur to him that such an undertaking could have important scientific aspects. He left Dana and the other scientists in his expedition behind in Australia, on the grounds that a scholar was 'worse than useless'.

Wilkes may have represented the general American attitude towards science, but it was an attitude that changed markedly in 1853, when Congress appropriated the necessary sums for the planning and building of a transcontinental railroad. Geologists were needed to determine the best routes for the railroad. A British millionaire and scholar, James Smithson, the illegitimate son of the Duke of Northumberland, had shortly before bequeathed half a million dollars to the United States for the founding of a scientific institute. One of the first great tasks undertaken by the Smithsonian Institution was investigation of the geological conditions in the area of future railroad lines.

For four years Dana conducted these researches in the West. The practical results, especially the great discoveries of mineral resources, gradually overcame prejudices against science. When a geologist discovered gold, and later oil, he could no longer be regarded as an unworldly fool. Dana took advantage of this shift in opinion. He worked up the facts he had gathered in the West and published the first fundamental work on the geology of America.

It was a fascinating and fantastic world to which Dana introduced his readers. He described the salt lakes and clay flats, the canyons and cliffs, the wild and largely virgin Rocky Mountains, the mud volcanoes and geysers, the forests of giant trees, the wastes of the Badlands, and the great lodes of gold, silver, copper, and petroleum – the unlimited riches and possibilities of the vast plains and mountains of the West, where only Indians and herds of buffalo roamed.

Much of what he saw seemed so much like the world in its primordial state that he almost inevitably began looking for fossils. Among other things, he found trilobites, extremely ancient marine arthropods from which spiders and crabs, and perhaps also the insects, have descended. But he did not explain them as old Timothy Conrad would have done, as proof of the Biblical Deluge. His orthodox contemporaries seem to have taken that amiss. For in his notes he set down a sentence that sounded like a sigh: 'Give trilobites of antiquity a tongue to speak, and they will correct many a false dogma in the theological systems.'

Dana made Yale University a centre of geological research and scientific progress. He stood up strongly against the politicians, pedagogues, and preachers who tried to banish the new insights of geology and biology from the schools and universities. Thus Dana almost inevitably came into contact with Darwin's doctrines.

The first prominent adherent of the theory of evolution in the United States was Asa Gray, who taught at Harvard University and for many years conducted a correspondence with Darwin. But Gray kept secret his 'heretical conclusion' that there are no independently created species. Only after Darwin's work on the origin of species was published did he come forward as an advocate of evolution. He arranged to have Darwin's book published in America. It was this edition Dana read, at first with considerable distrust; he had hitherto believed in Cuvier's theories and also felt his religious faith offended: he was, after all, a believing Christian. But eventually Dana got in touch with Asa Gray, and himself began corresponding with Darwin. Some time later, after passing through a period of thorough and honest self-examination, he became a convert to the theory of evolution. Thus the struggle over Darwinism was opened in America. It led to near riots, trials, and court verdicts. But it also shaped the men who established the age of science in America.

'With God Against Evolution!'

Ever since the time of Darwin, American scientists have repeatedly endeavoured to reconcile the theory of evolution with the Christian religion. Dana took the lead; he was the first prominent scientist in the United States to proclaim that evolution did not make God superfluous, but was in fact directed by God and was a manifestation of divine power and grandeur. Many American clerics held the same opinion. Henry Ward Beecher, the most popular evangelist of the day, preached that the theory of evolution was the key to many secrets of nature and a constant revelation to man of the works of God.

Dana went further than most in including man within the scheme of evolution, even as Darwin was doing. The evolution of man is likewise subject to the intervention of a higher power, he declared. But he did not hesitate to state his belief that man was descended from the apes.

Thus a brand of Darwinism with religious coloration entered the precincts of Yale University. But to many Americans this kind of evolutionary theory was in itself sacrilegious: such famous contemporaries of Dana as Matthew Fontaine Maury, the founder of scientific oceanography, proclaimed that the Bible was *the* authority on questions of science. At Harvard the reaction was even stronger. Despite the presence there of Asa Gray, Harvard became a fortress of anti-Darwinism in America. For the most famous scientist at Harvard was also America's most vigorous anti-Darwinist. That was Louis Agassiz. But whereas Agassiz fought his battle with Darwinism on an objective and scientific plane, other scholars resorted to invective. The poet James Russell Lowell made 'with God against evolution' his battle cry. Lowell labelled the theory of evolution a stew that would be a poor substitute for the rock of eternity.

Jonah and the Sea Serpent

Around 1860, when most American university professors were as yet more inclined to believe in Jonah's whale than in slow and gradual change of the earth and its organisms, some discoveries were made that initiated the great epoch of palaeontological research in North America.

A citizen of Philadelphia, William Parker Foulke, came upon the skeletal remains of a monstrous animal in a marl pit in the suburb of Haddonfield. Some of the bones had already been carried away as souvenirs, but enough remained to arouse Mr Foulke's interest. He had them moved to the Philadelphia Academy of Natural Sciences for closer examination.

The professor of anatomy who studied the huge bones was Joseph Leidy, a secret adherent of Darwinism. He was considered the foremost authority on fossil vertebrates in America. He speedily recognized that the Haddonfield bones were those of a giant saurian akin to the iguanodon. This *Hadrosaurus*, a giant saurian with a peculiar duck-like beak, was the first of the many dinosaurs to be found in America.

Similar saurians of the Cretaceous period were subsequently discovered in the American deserts in what seemed a mummified condition. It looked as if the entire skin, down to the finest details and most subtle contours, had been preserved. Even the muscles and sinews seemed to be still present. In reality these specimens were pseudo-mummies. Fine, dry, powdery sand had once covered the cadavers and filled in all the folds in the skin. In the course of the decay process the sand everywhere replaced organic substances. Gradually it hardened, became sandstone, and the stone was a perfect cast of the cadaver.

Leidy applied himself with great eagerness to James Dwight Dana's works on geology, to learn which American formations were most likely to contain more such monsters. Whenever his time permitted, he went digging on his own. In his Academy lectures, he often declared that the West, once it had been opened up, would become an Eldorado for palaeontologists.

A twenty-year-old student from a wealthy Quaker family heard those words and remembered them. Eleven years later this man went west and fully confirmed Leidy's prophecies. His name was Edward Drinker Cope.

Dr. Koch's Forgeries

According to family annals, Edward Cope was only six years old when he began taking an interest in fossils. It is at any rate a matter of record that as a small boy he was greatly excited by a remarkable fossil then being exhibited in Philadelphia, the skeleton of the *Hydrarchus*, which was so long that it stretched through three rooms.

This 'hydrarchus', the sea serpent of sailors' yarns, happened to be a swindle, a palaeontological forgery which appealed to the credulity and sensationalism of the masses. Around 1840 a large number of whole and fragmentary skeletons were found in Alabama. Richard Harlan, the Philadelphia physician who had tried to classify the fossils of North America, took the bones to be those of a giant reptile and gave it the proud name of *Basilosaurus*, 'Emperor of the Saurians'. But Richard Owen in London soon determined that it was not a saurian, but a large, predatory whale of the Early Tertiary, *Zeuglodon*.

Word of Owen's reclassification was slow to reach America. The primordial whale with its gigantic maw, huge teeth, and enormous skull presented so fearsome a sight that a profit-minded German immigrant, one Albert Koch who called himself a doctor, hit on the idea of buying up some of these bones and putting them on show.

Koch had experience in such matters. He had once before constructed a monster out of several elephant and mastodon skeletons, which he exhibited in England and America, charging a small admission. He called it the *Missourium*, the 'Missouri animal', and even so well trained a palaeontologist as the discoverer of the iguanodon, Gideon Mantell, fell for the swindle and wrote enthusiastically: 'It is the largest of all hitherto known fossil mammals – thirty feet long and fifteen feet in height.' Naturally, the British Museum bought the creature. When the laboratory workers at the museum realized the deception, they had great difficulty in taking Dr Koch's absurd construction apart again. Nevertheless, they finally succeeded in assembling a fairly complete mastodon out of the bones.

Before long Koch had once again put together a giant skeleton out of the bones of several ancient whales; he took it to Europe and presented it to Frederick William IV of Prussia as the 'behemoth of the Bible'. The king was ravished by the 'behemoth'; instead of buying it, he ordered that Koch be paid an annual pension of one thousand imperial talers. At this time the Prussian king was already suffering from softening of the brain, which probably made the deal easier. The behemoth was exhibited in the Berlin Museum of Natural History. There the scientists on the staff soon realized the forgery, but the king would not hear of withdrawing his sinecure.

Since Koch had had such good luck with the Missourium and the behemoth, he now set about constructing a hundred-foot sea serpent out of another stock of ancient whale bones. It looked exactly like the sea serpent of nautical yarns: with raised head, wide-open maw, vigorous fins, and a handsomely undulating body. Koch called it *Hydrarchus sillimani* in honour of Dana's father-in-law. Benjamin Silliman could not object to the misuse of his name, since he was no longer among the living at this time.

Since the scientists of the British and Berlin Museums were doing all they could to denounce him, Koch did not linger in Europe with his sea serpent. He showed it in Dresden, Breslau, Prague, and Vienna, but only briefly, for no sooner did he appear in a city than the indignant Gideon Mantell was warning collectors and museum curators against this damnable swindler.

In America, Koch had a similar experience. For a while he was able to show the sea serpent in New York. But soon he had to pack it hurriedly into its many crates. A professional anatomist and zoologist, Jeffries Wyman, had visited the show and demonstrated in a learned article that the skeleton of the alleged hydrarchus could in the first place not have come from a reptile and in the second place not from a single individual.

Mantell, too, wrote from England and warned his colleagues in New York that the sea serpent was just as much of a cynical forgery as the Missourium by which he had been hoodwinked.

Koch had no desire to become involved in detailed discussions with Professor Wyman. He also wanted to escape the stubborn pursuit of Dr Mantell. He therefore turned his back on New York and went on tour with his sea serpent to less unpleasant regions, where people were grateful for a bit of fantasy. Finally he sold the monster to the Wood Museum in Chicago, where it met its end in the great fire of 1871.

But the zeuglodon which Koch had raised to such dubious fame subsequently became one of the chief testimonies to the descent of whales from carnivorous land mammals. And Edward Drinker Cope, who as a small boy had gaped at the fabulous sea serpent, twenty-five years later discovered real sea serpents in the Chalk of Kansas – mighty snake-like mosasaurs and tylosaurs which must have looked a good deal like the fabulous monster of seamen's yarns.

An American Lamarck

Edward Drinker Cope was a born scientist. Like Cuvier and Darwin, he had a flair for scientific insights. At the age of ten he was making excellent drawings of the ichthyosaurs whose plaster casts he had seen in the Philadelphia museum. His teachers urged him to study anatomy. His prosperous father, however, expected him to take up agriculture and help administer the family estate.

When Cope was eighteen years old, his father tried to lure him from his naturalist's hobbies by giving him 'McShag's Pinnacle', a fine, profitable farm which was part of the family's vast holdings. But young Cope was interested only in the birds, snakes, and fish that he could observe on the farm; tilling the fields and raising livestock had no attraction for him. There were a number of disagreements with his father, a hardheaded Quaker who regarded zoology as no sort of vocation. But Cope finally won permission to attend the University of Pennsylvania, where he began by studying anatomy under Professor Joseph Leidy.

During his first semester he read with intense interest Darwin's *Voyage of the Beagle*, and then the *Origin of Species*. Darwin's theory was being violently criticized by most of his teachers and fellow students. But since his revered teacher, Professor Leidy, regarded the Darwinian theory as highly suggestive, Cope resolved to study the phenomenon of evolution as thoroughly as possible. He became particularly interested in aspects of evolution that could be revealed by fossil vertebrates.

Soon circumstances favoured his pursuit of this interest. When the Civil War broke out, his father was anxious to keep his son from being drafted for military service, since that ran counter to Quaker principles. He therefore sent Edward to Europe for further study, and provided him generously with money.

In Europe young Cope visited almost all the noted universities and natural history collections. He went to Paris, London, Berlin, Munich, Leiden, and Vienna. What he heard among the followers of Darwin, Huxley, Haeckel, and the other biological revolutionaries led him to profess the theory of evolution without the reservations his countrymen Gray and Dana had felt compelled to make.

At the age of twenty-four Cope returned from Europe carrying in his intellectual baggage not only the ideas of Darwin, but those of Lamarck as well, which he had become acquainted with in Paris. Upon his return he at once obtained a professorship in zoology at the then small and unpretentious Haverford College in Pennsylvania. He married his cousin Annie Pim, and devoted himself to family life and in particular to his beloved small daughter Julia. Teaching at the modest college did not really fulfil him. But he was still young and seemed to feel that he had time. It was therefore something of a shock to his family when he suddenly resigned his professorship, sold the farm, 'McShag's Pinnacle', and set out for the great open spaces of the West to hunt for fossils.

The Passionate Collector

Cope chose Fort Wallace in Kansas as the starting point for expeditions and excavations. With two covered wagons, seven companions, and fourteen mules, he travelled over prairies inhabited mostly by roving Indians and wild animals. For the small party of scientists, the advance into the wilderness was fraught with peril. The Indians, forced out of their time-honoured hunting grounds, were engaged in a perpetual guerrilla war with the palefaces. But Cope refused to be frightened. He got in touch with the troops who were guarding the West, informed them of his plans, and persuaded them to keep a lookout for fossils. Soldiers and colonists, trappers and prospectors were promised good rewards for every find.

Within a short time Cope had such striking successes that at once he became the most famous palaeontologist of the age. He dug out of the Kansan Cretaceous those 'sea serpents' that so strongly reminded him of his childhood and Doctor Koch's fraudulent monster. These mosasaurs, *Liodon dyspelor* and other species, were gigantic American relatives of the Petersberg monster Cuvier had once upon a time stolen from Canon Godin. Long-necked plesiosaurs were soon added to the collection, among them *Elasmosaurus platyurus*. In addition Cope found a wide variety of other land saurians and flying saurians. It was clear that North America during the Mesozoic had been a peerless saurian paradise.

In the prairies, mountain valleys, the river beds of Kansas and Wyoming he also turned up many other fossils of a different sort: trilobites, fishes, giant tortoises, and above all a number of bizarre mammalian creatures of the Early Tertiary. Professor Leidy had been right: here in the West the stages of American palaeontological history were coming to light with amazing completeness.

Cope kept his finds in two adjoining houses he had bought on Pine Street in Philadelphia. Before long, a gigantic storehouse of bones accumulated in these houses. Cope had gathered such immense quantities of fossils that he could not possibly reconstruct and provide exact descriptions of even half of them. For the present, however, he turned none of them over to American museums and collections; in fact, he continued to enlarge his stock by buying up great collections from other countries.

When a particularly rich collection of large mammals from the South American pampas was shown at the Paris World Exhibition, Cope immediately sailed for Paris and bought up the giant sloths, glyptodons, toxodons, and other archaic creatures which had once so fascinated Darwin. Meanwhile he had inherited a large fortune from his father; he felt he could give free rein to his hobby and possess a private collection which would be unique in the world.

His collecting passion became such a mania that he soon lost all view of his own treasures. Dust-covered crates filled with fossils stood around in corners, still unopened and unexamined. Only after his death was it discovered what treasures they contained.

Sometimes passers-by would stop at his houses on Pine Street to peer through the windows at the weird skulls and skeletons piled up inside. Among these relics of long extinct species crawled living animals, such as Cope's pet turtle. During the winter months, when he could not dig in the West, the owner of these treasures could be seen prowling among his bones – a high-strung man with a Vandyke beard, forever classifying, writing up scientific reports, and trying to recognize in his fossils the ancestors of living forms.

Cope's Bible

A good many ideas were born among those heaps of bones on Pine Street. Cope had begun publishing some of his theories in 1871, when he was just thirty-one years old. His starting point was Darwin's view that the fittest must survive the struggle for existence. Cope kept turning this over in his mind. He asked himself: 'Where do the fit variations come from? What is their origin? Which causes produce fit variations?'

Thoughts of this sort led Cope back to the theory of Lamarck, which had been forgotten for almost half a century. Darwin's law only represented a limitation, he concluded, preserving or destroying something that already existed. But there was still the need to examine the laws by which those forms had come into being. In other words, Cope wanted to supplement Darwin's 'survival of the fittest' with a discussion of the 'origin of the fittest'. The factors that created useful varieties must be sought in the environment, he believed. He concluded that it was not selection, but such environmental conditions as climate, food, the use or disuse of organs, that changed species. Following Lamarck, he decided that if a given limb had great demands placed upon it in locomotion, it increased in length, and such an acquired characteristic would be inherited. A large number of creative forces were active in nature, he believed, producing changes in species. He gave these factors difficult and often jaw-breaking names: ergogenesis, stratogenesis, bathmogenesis, mnemogenesis, emphytogenesis, and so on. Thus he became the founder of a complicated, somewhat incomprehensible but much-discussed Neo-Lamarckism in America.

Cope's theories of evolution are given short shrift nowadays; the discovery of the mechanisms of genetics and mutation outmoded them. His scientific and historical importance lies in

SOUTH AMERICA

Caenolestine marsupial

Marsupial carnivore·

Camel-like litoptern·

Horse-like litoptern

Toxodont

Homalodothere

NORTH AMERICA

Shrew

Wolf

Camel

Horse

Rhinoceros

Chalicothere

Some convergent types among North and South American mammals, mostly extinct forms. All are drawn to the same scale.

another field entirely. Cope's friend and successor, Henry Fairfield Osborn, once made the point that Othniel Charles Marsh, Cope's rival, always knew where to look for interesting finds, but that Cope had perceived that the real work began after the find had been made.

As a result of his studies, so vast that he was never destined to complete them, Cope came to be the greatest comparative vertebrate anatomist of the age in America. His book on the fauna of the American Tertiary was the standard work in the field; it was called by friends and enemies, with a mixture of admiration and sarcasm, 'Cope's Bible'. He only half finished it. It was Cope's misfortune that he squandered much of his strength, which he needed to carry out his scientific tasks, in a personal struggle. For Cope was not the only fossil hunter to be roving through the Wild West during those years and bringing bones back home to be catalogued and ordered. Soon after his arrival in Kansas he encountered Othniel Charles Marsh, who was rapidly to become his formidable rival and bitter enemy.

'And He Went Out to War'

In the Book of Judges there is a figure named Othniel of whom it is said: 'And the Spirit of the Lord came upon him, and he judged Israel, and went out to war.' It had always been the custom of the Marsh family to give their children Old Testament names. Caleb Marsh gave one of his sons the rather unusual name of Othniel probably because the two names appear in the same passage – although in the Bible, Caleb is Othniel's younger brother.

Whatever the reason, the name was most apt for the personality. For the spirit indeed came over Othniel Charles Marsh, the farmer's son from Lockport, New York. Moreover, he fought his way up to become a judge in his science, and he certainly 'went out to war'. It was a war that dominated a quarter of a century of American palaeontological research and ultimately became a national scandal.

Othniel Charles Marsh enjoyed enormous prestige in the scientific world and amassed a unique collection of dinosaurs, pterodons, and giant mammals. Nevertheless, he had not enjoyed a good press with his biographers. He was a rich, highly gifted but also obstinate and sometimes brutal man, one of those pioneer types who helped shape the destiny of America. He had far larger funds at his disposal than his rival Cope. This fact was due to chance, a chance that proved to be the decisive element in his life.

His father Caleb, a strict and narrow-minded man, had married the sister of a poor street peddler named George Peabody. She died soon after Othniel's birth. But her brother George rose from peddler to storekeeper and eventually to entrepreneur, banker, and multimillionaire; his life story reads like old-fashioned schoolbook versions of the careers of Edison and Rockefeller. For a time he was one of the richest men in America. The world-famous Wall Street house of J. Pierpont Morgan sprang from Peabody's banking enterprises.

Since Peabody had no children, he devoted a large part of his fortune to philanthropic purposes. One of his friends was the British manufacturer Robert Owen, a socialist idealist who like himself had risen from simple circumstances to the possession of millions. Back in England, Owen had endeavoured in vain to transform his textile factory into a model collective enterprise which would be the nucleus of a society based on equality and justice. During a lengthy stay in the United States, Owen established the communistic community of New Harmony in Indiana, in the hope that compulsory education, the creation of favourable social conditions, and the establishment of co-operatives would be a cure for social ills. Peabody supported the experiment and continued to provide for New Harmony after Owen had returned to England. In addition Peabody gave heavy endowments to model colonies for the poor in England and America, for the establishment of schools, and for the education of Negroes in the South.

Naturally brother-in-law Caleb Marsh hoped for some share in these riches. But Peabody preferred his young nephew Othniel, who reminded him poignantly of his lost sister. He cherished the boy, took charge of his education, and encouraged his desire to be a naturalist. Perhaps he did so out of a certain spite against the boy's father, for old Caleb regarded all science as humbug. But the crabbed farmer was no match for the boy's fabulously wealthy uncle.

Like Cope, Marsh too had made his first acquaintance with fossils as a child. During the digging of the Erie Canal, which passed close by his father's farm, trilobites and other fossil marine animals had been found. These strange petrified creatures fascinated Othniel; he began systematically collecting them, and while still a schoolboy entered on a correspondence with Agassiz, who gave him the idea of studying palaeontology. Uncle George Peabody agreed, and recommended that he acquire the necessary learning in Harvard.

But Marsh did not go to Harvard, although Agassiz was teaching there. Instead he chose the then far less important

Yale University. This decision was well considered from a tactical point of view and proved that the youthful student Othniel was even then unusually foresighted. In venerable Harvard with its established traditions, its world-famous scientists, and with Louis Agassiz towering above all, Marsh would have had to wait years to rise to the scientific heights to which he aspired, even if Uncle Peabody's millions were there to aid him. In developing Yale, on the other hand, where Dana was teaching the new geology and where Lyell, Darwin, and other scientific rebels were not banned, any young talent had greater chances.

While at Yale, Marsh soon made another clever move. Agassiz had asked him to coax his uncle into making a large contribution to science at Harvard. Peabody agreed. But Marsh now saw to it that the promised gift was diverted to Yale rather than to Harvard. This 'betrayal' naturally strengthened Marsh's ties with Dana and clouded his relationship to Agassiz. And it also engendered tensions between Yale and Harvard that were a good deal more serious than the subsequent athletic and scientific rivalries. But Marsh did not especially worry about that. He was concerned with rising as quickly as possible to the role of a leading Yale man. As one of his biographers, Bernard Jaffe, has written, he carried out this plan with the obstinacy, single-mindedness, zest, and ruthlessness of an industrial magnate. He meant to devote himself to the service of science on the colossal scale that Peabody's millions permitted.

The Peabody millions flowing into New Haven indirectly helped to make Marsh a Yale professor. He thus became the first professor of palaeontology in the United States, for Leidy, Dana, and Agassiz taught the subject only as an auxiliary matter. With Peabody's money Marsh founded the Peabody Museum of Natural History in New Haven – the first great palaeontological museum in North America. Whatever he laid hands on he was determined would be bigger, better, and more complete than anything that had existed before.

From 1863 to 1865 Marsh, like Cope, studied at various German and English universities. Thus he too missed the Civil War in his homeland and encountered instead the war raging over Darwinism in the Old World. Abroad, all doors were thrown open to George Peabody's nephew. He met Lyell, Darwin, and

Huxley; he spent his money on fossils, which he bought up by the ton at many European sites and had shipped to America. Thus his museum in New Haven soon had foundation stock on which to build.

Cope's European studies and impressions had, as we have seen, made him an adherent of Neo-Lamarckism. Marsh, as a consequence of his talks with Darwin and his friends, became a fanatical Darwinist. In his robust way he ignored the fact that the concept of evolution was wisely held in odium in the States. 'To doubt evolution is to doubt science,' he told his fellow countrymen upon his return, 'and science is only another name for truth.' This kind of blunt talk went down better with Americans than all the preceding discussions. Marsh became Darwin's 'American bulldog'.

Buffalo Hunters in Indian Territory

Since Marsh had ample funds at his disposal, he could draw up a far more ambitious plan for the palaeontological conquest of the West than Cope. He decided on a systematic investigation of eight states of the Middle West and Northwest, and enlisted in his cause not only the scientific apparatus of Yale University, but also the railroad companies, the army, and even the Indians. The building of railroads across the continent had meanwhile begun. With the penetration of the West went the notorious slaughter of sixty million American bisons, the buffalo of the prairies. With it also went the bloody Indian wars.

Marsh was involved in all these developments. For a number of years he undertook regular summer journeys with professors and students of Yale and a number of interested businessmen. These expeditions yielded, among other things, the numerous fossil horses which were to make Marsh's name famous the world over. After a while he began to play a significant part in the fate of the Indians.

At first Marsh was able to dig in Indian territories only under difficult conditions and often in peril of his life. For the Indians, especially the Sioux, were roused to fury by the invasion of their lands by prospectors and adventurers. Knowing nothing about palaeontology, they took every scientist who worked with pick and shovel to be a prospector, and were out to kill him.

It must be granted that Marsh and his adventurous companions were not entirely innocent in their relationship with the Indians. The great fossil hunter did not give the impression of being a quiet scientist. He was a first-rate rider, hunter, and shot, precisely the kind of frontiersman the Indians feared and had learned to fight. Thanks to his influence, Marsh obtained armed troops to accompany him into the Indian territories; and the Sioux were only too familiar with the ruthlessness of these U.S. Army units. Moreover, Marsh was on the most cordial terms with the Indians' arch-enemies, the men of the Union Pacific Railway Company. They supported him wherever they were able.

The guides and scouts whom Marsh employed for his expeditions likewise did everything to enrage the Indians. Among them was William Cody, the celebrated Buffalo Bill, the king of bison hunters. His example infected the palaeontologists. For recreation, Marsh would jump into the saddle and dash off with his rifle into the midst of the buffalo herds. In the course of one such hunt he fell from his horse and came within an ace of being trampled under the hoofs of the herd. A buffalo hunter who was partly responsible for the spread of famine among the Indians was naturally not regarded as a welcome guest by the Sioux.

The tensions in the Indian territories finally increased to such a point that it seemed as if further expeditions would have to be renounced. But it was just at this crucial point that Marsh had an experience which made him the friend and advocate of the Indians. Linked with this story were fossils of a particularly fascinating kind – the 'thunder beasts' of the Indian legends.

The Mystery of the Thunder Beasts

The Great Plains of South Dakota, Nebraska, and Wyoming merge in places into barren hills almost without vegetation. In prehistoric times torrential rains cut deep cracks and furrows into the land. Subsequently, prolonged drought and the searing summer sun turned these areas into stony deserts. But occasional cloudbursts wash out layers of soft clay and bring to the surface bones that testify to the onetime teeming life of these regions.

The first French explorers of these areas called them, with open repugnance, *Les Mauvaises Terres*. The American pioneers translated the name to Big Bad Lands. There, after bloody wars and expulsions, the last of the Sioux tribes took refuge. There too the last of the buffalo herds sought asylum.

But the pioneers in their covered wagons, the prospectors, and the railroad workers, soon reached the Badlands also; and with them came alcohol, smallpox, tuberculosis, and the other vices and plagues of the palefaces. Year after year a million buffaloes were killed. Thus the Sioux were deprived of the basis of their sustenance. Their dissatisfaction grew steadily. The last great Indian uprising was developing in the Badlands.

One of the few whites who enjoyed the confidence of the Indians in this critical situation was Captain James Cook, last of the famous scouts. He lived in northwestern Nebraska as a trapper, traded with the Sioux, and took some interest in their cults, mysteries, and legends. In the course of his dealings with them he repeatedly came upon the legend of the thunder beast.

The mythology of the thunder beast occupied a prominent place in the religious and everyday life of the Oglala Sioux. It was associated with the buffalo cult and thus with the dependency of the Indians on their most important hunting and totem animal. The Oglala cherished the curious belief that during the great thunderstorms that regularly descended upon the prairie gigantic animals leaped out of the clouds. These 'thunder beasts' or 'thunder horses' would help the Sioux in times of famine by driving herds of buffalo towards them. After the storm was over, the beasts disappeared into the soil of the prairie.

Once when Cook was talking with Oglala Chief Red Cloud about the thunder beast legend, Red Cloud beckoned to his underchiefs, Little Wound and American Horse. The two Indians led Cook to a hiding place in the camp and showed him a huge jawbone with several molars still in it. This bone, they explained, came from a thunder beast. After every big storm such bones were washed out of the ground, they declared.

Cook had no idea what kind of animal the huge creature was. Since he was more interested in living Indians than in weather-beaten bones, he paid no further attention to the fossil. Possibly the myth of the thunder beasts would never have reached the public, and would gradually have died out with the passing of the Sioux tribal life, if Cook had not met Othniel Charles Marsh around 1875. The contact between these two very different men led to the discovery of the magnificent Early Tertiary fauna of the Great Plains.

Politics and Palaeontology

Marsh informed Cook that because of the growing tensions in the Indian territories, he could hardly continue his research work. Cook, for his part, told Marsh about the thunder beasts of the Oglala, and offered to act as intermediary. Marsh, naturally, was extremely interested both in the thunder beasts and in good relations with the Indians. He accompanied the captain to the Oglala reservation and had a long talk with Chief Red Cloud.

From Cook and Red Cloud, Marsh learned a great deal about the misery of the Indians, the mistreatment they were receiving the corruption of the government offices involved in dealings with them, the constant theft of their land, the countless breaches of treaties, and the crimes of the prospectors, buffalo hunters, and Indian agents. He realized that the Badlands might soon be one vast battlefield unless the government could be prevailed on to change its entire Indian policy.

What Marsh learned in the Black Hills, on the border of South Dakota and Wyoming, transformed him into a passionate defender of Indian rights. Ambition may have made him unscrupulous towards his rivals, but where the Indians were concerned, he felt and acted from genuine warmth. The turning point was that conversation with Red Cloud. Marsh was one of the first Americans of distinction to throw the entire weight of his personality into the scales to help the Indians and influence public opinion in their favour.

From the Black Hills, Marsh travelled straight to Washington, called on President Grant, and categorically demanded that he dismiss everyone responsible for the bad treatment of the Indians, from the pettiest civil servant up to the Secretary of the Interior, replacing them by honourable men well disposed towards the Indians. So greatly had the prestige of scientists increased in the United States since the Civil War that Marsh was amazingly successful. He succeeded in obtaining the resignation of the Secretary of the Interior and the dismissal of a number of corrupt or incompetent government officials – and on top of all won acclaim for having battled the Goliath of government. The common people did not cheer Marsh's accomplishment out of any newborn love for the Indians. They simply felt a deep satisfaction at his having embarrassed the Grant administration. President Grant had been making himself more and more unpopular from day to day. His times were not only one of pioneering conquests in the West, but also an era of trust-building, ferocious competition among railroad, meat, steel, and oil magnates, and of an extraordinary orgy of speculation, political manoeuvring, and corruption. Marsh had lanced one of the suppurating sores in the American body politic, and all the discontented regarded him as a hero.

To Marsh's disappointment, the Indians experienced no significant improvement in their situation. For before President Grant could embark on the necessary reforms General Custer, the commander of the troops in South Dakota, destroyed all prospects for a new orientation by his clumsy policies. New gold finds had been made in the Black Hills at just this time. Custer let news of the discovery get out, thus attracting a fresh swarm of prospectors to the country. He supported the prospectors when they attempted to take possession of the Indian reservations in the Black Hills, and assured them armed protection. The ultimate result was that bloody Sioux uprising of 1876, in the course of which General Custer and his men were wiped out by the Indians.

Marsh and his fossil collectors remained entirely unaffected by these events. The Sioux now knew that the Yale professor was their friend and champion. They permitted him to dig undisturbed in their territory, and even adopted him into their tribe, giving him the honorary name of Wiscasa Pahi Huhu, 'the man who chops out bones'.

The Titanotheres' Horns

The fossil that Red Cloud had shown Captain Cook was baptized by Marsh *Brontotherium*, 'thunder beast'. Marsh, his associates, and a number of later palaeontologists, among them Osborn and Scott, subsequently discovered a whole series of different thunder beasts. Marsh called the entire group *Titanotheridae*. They were mammals, odd-toed, related to the rhinoceroses, tapirs, and horses. Small types, barely the size of sheep, had appeared in North America in the Eocene. In the Oligocene they developed into huge animals about halfway between a rhinoceros and an elephant in size.

The light-green clay under the deserts of Dakota, Nebraska, and Wyoming had so faithfully preserved the shapes of the titanotheres that Marsh's associates could even venture to reconstruct their musculature. In so doing they inaugurated a wholly new branch of palaeontology: the effort to make determinations on the soft parts of fossils and on the functions of the organs. Bones are comparatively inactive, as Sherwood L. Washburn has remarked; the muscles are far more important for the functioning of the organism. Muscles can influence the form of the bones, and Washburn postulates that relatively minor genetic changes in the musculature may have led to rapid and fundamental transformations of the entire organism.

In the case of the titanotheres, however, the evolution of their striking skull adornment proved even more instructive to biologists than the reconstructions of the muscles. For the development of the skull provided the scientists with a persuasive example of the meaningful interplay of adaptation and selection.

Among the oldest types of brontotheres, from the Early Tertiary, the skull was marked only by a rather unobtrusive thickening of the skull bone which seemed to have no purpose. Nevertheless George Gaylord Simpson, the American palaeontologist, commented in 1944 that even at this stage the animals probably had the habit of ramming their rivals and enemies with their heads. When they grew into strong, stocky creatures with thick skulls, they had practically no other weapons at their disposal. A heavier bone in the region of the skull, with which they were accustomed to butt, was therefore of advantage to them.

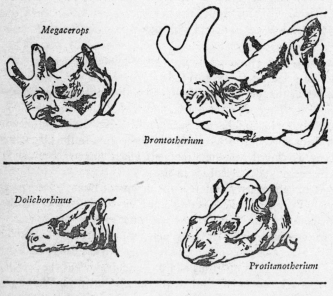

Megacerops

Brontotherium

Dolichorhinus

Protitanotherium

Eotitanops

Development of the brontothere's skull, a favourite example of purposeful evolution.

The seemingly pointless and accidental thickening of the skull proved its value in the struggle for existence. Consequently the evolutionary trend continued along the same lines; the bumps on the head grew large, more fitted to their function. Within a short time they were extremely effective weapons. The animal continued to increase in size, and the horns to enlarge, until at last there was a row of spikes growing out of the nose, somewhat different in each species. The brontothere proper, for example, bore a huge bony knob, probably covered with skin, on its nose.

The example of the brontotheres might be extended to many other groups of animals. The smallest advantageous characteristic can suddenly prove surprisingly effective and thus decisively influence the course of evolution. For animals repeatedly seek to explore their available environments and to occupy these to the best of their abilities. As soon as some at first insignificant trait is useful to them in this effort, it is augmented and perfected by the cooperative working of adaptation and selection.

A Tertiary Tsetse Fly

The brontotheres of the Great Plains, like many other groups of North and South American animals, belong among those New World mammals which after a longer or shorter period of florescence vanished utterly, leaving no descendants. Whole generations of scientists have explored the problem of the curious rise and decline of the Great Plains fauna. By now a good deal is known about them. Huge herds of brontotheres inhabited the flood plains of the great rivers, living like elks on leaves and buds. With their armoured skulls, they had no significant enemies. They were scarcely threatened by volcanic eruptions or catastrophic floods.

Most palaeontologists therefore attribute their extinction to the change of climate during the transition from the Early to the Middle Tertiary, which gradually transformed wet areas into parched steppes. The brontotheres' teeth, they reason, were adapted only to the chewing of leaves and soft plants, not for grinding up hard steppe grasses. This happens to have been the case for all Early Tertiary ungulates. They all lived on foliage. When the vegetation changed, some species of hoofed animals managed in a variety of ways to adapt their dentition to the new conditions. The elephants, the even-toed ungulates, the rhinoceroses, and the horses succeeded in doing so. Why did the brontotheres fail?

Henry Fairfield Osborn, the pupil and successor of Cope, developed an interesting and much-discussed theory to account for the extinction of these and other animals. The brontotheres, he believed, had persistently continued along the line of specialization. They had adapted all their qualities to the ultimate. For a time their dentition was highly useful in the damp forests along the riverbanks, but under other conditions such dentition would have been faulty. Since it was useful in the beginning, it developed more and more into a specialized leaf-eating dentition. The disaster came when the brontotheres were forced to shift to grass food. Then the hitherto useful molars wore down so quickly that the animals could not hold their own in the struggle for existence.

But the thunder beasts might have come to their end from quite other causes. In the same Oligocene strata in which the brontotheres' bones are imbedded, American palaeontologists have discovered a prehistoric fly, *Glossina oligocenica*, a close relation of the present-day tsetse fly. The dry steppes which replaced the damp woods of Oligocene North America must have greatly favoured the multiplication and spread of this insect. Perhaps it can be assumed that the fly was a carrier of trypanosomes, which are closely related to the agents of the cattle plague, the sleeping sickness, the nagana and surra diseases of today.

During the Cenozoic era tsetse flies made vast regions of the earth uninhabitable for man and animal – especially after the introduction of domestic animals, which, unlike many African wild animals, are not immune to the disease spread by the fly and therefore are chiefly responsible for its dissemination. If the white man had not virtually wiped out the wild game of Africa, but instead protected it and used it for the nourishment of the population, the epidemics caused by the trypanosomes could not have spread with the virulence they had after wild herds were replaced by herds of domestic cattle. It may be assumed that constant selection had made the wild animals immune, and that in the course of this process of selection a good many species were wiped out because they could not withstand the tsetse fly.

The same sort of development may have taken place in the Oligocene in North America. Possibly the brontotheres were particularly prone to trypanosome diseases. They had developed no resistance to these diseases in their wooded flood plains. When the land parched, they had their first encounters with tsetse flies. If a species cannot mobilize defences against a new epidemic, it may be brought to the verge of extinction, and only a little shove will push it over the brink.

All this, of course, is merely hypothesis with a more or less high degree of probability. In Marsh's day no one gave much thought to the riddle of extinction. The palaeontologists of the time were chiefly interested in collecting the largest possible number of finds.

Marsh's successes in this regard were so considerable that he soon could claim a virtual monopoly of all the fossils in the western states. He did everything he could to exclude all potential rivals from the area. Thus the fossil hunting in the West turned into the famous Battle of Bones.

The Dawn Emperor

When Cope turned up in Fort Wallace and began excavating his first fossils, Marsh had just finished off another rival. This was Professor Arnold Guyot of Princeton, a conservative scientist who clung rigidly to the dogma of the immutability of species.

Guyot was in the habit of purchasing fossils that had fallen into the hands of Kansan and Wyoming farmers and frontiersmen. He was not an especially brilliant man, but his attacks on the theory of evolution vexed a convinced Darwinist like Marsh. It is understandable, therefore, that Marsh decided to put a spoke in Professor Guyot's wheel. He did so by a trick that was to prove effective many times in the future: he offered the people who were working for the Princeton professor twice as much money if they would collect for him. To show them how seriously he meant his offer, he would descend on them in a private train, and pay cash down for whatever they had to sell. Guyot could not compete with means like this, and had to give up his fossil project.

But Cope was quite another sort of adversary. Marsh branded his Philadelphia rival a bungler, an unqualified intruder, a palaeontological poacher; but Cope's prestige was quite as great as Marsh's own. And Cope was able to repay Marsh in his own coin. He regarded Peabody's well-heeled nephew not as a serious scientist, but as an unusually egotistic, greedy collector, an exploiter and pillager of palaeontological sites. If the Yale professor raised his bids to the collectors, the Philadelphia Quaker's son kept pace with him. It was almost like an auction, and for some time no one could say which man would have the greater stamina or the larger resources. The people who profited most from all this were the farmers, prospectors, soldiers, schoolmasters, and railroad workers with wit enough to recognize and extract fossils.

In the course of the competition they were able to raise their finder's fees higher and higher.

A Garbled Telegram

Open war broke out between the two scientists when Cope discovered in the Wyoming Early Tertiary the first representatives of a hitherto unknown order of mammals. They were members of an extinct family of ungulates, like Marsh's brontotheres; but they had no parallels with any living or fossil group. Cope's first impression was of huge creatures combining the traits of rhinoceros, elephant, hippopotamus, and pig. He found the first of them along the Green River in Wyoming, and gave it the somewhat bombastic scientific name *Eobasileus*, 'dawn emperor'.

As Cope reconstructed him, the dawn emperor looked truly weird. He resembled an elephant in having a long trunk and pillar-like legs, but he was also graced by enormous horns and sabre-like tusks. Cope announced that Eobasileus was the most extraordinary fossil that had ever been discovered in America. Marsh took this claim in bad part, since it seemed to belittle his beloved brontotheres.

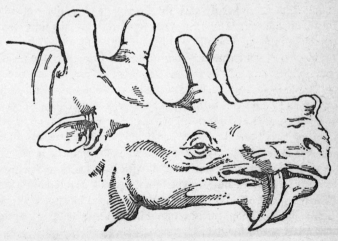

The 'dawn emperor' found by Cope in Wyoming.

Soon afterwards Cope came upon another fossil that obviously belonged to the same mammalian group. It looked somewhat different from the dawn emperor, but likewise seemed a creature sprung from the imagination of a Breughel or Hieronymus Bosch. When Cope unearthed this find, his nerves nearly gave way. He knew that Marsh and his staff of assistants were digging fairly nearby in the same Early Tertiary strata. It was therefore quite possible that he would come upon similar bones. If he used his connexions to name and publicize it more rapidly, Cope would lose his priority.

With this in mind Cope had made a practice of utilizing the telegraph service, whose wires had only recently reached these regions. As soon as he made a discovery, he would transmit the information to Philadelphia, giving the scientific name he had chosen, and the characteristics of the fossil. In this way, he thought, Marsh could not cheat him out of his priority.

He did so in this case. But this time he had made up too difficult a scientific name: *Loxolophodon*, an allusion to the animal's peculiar slanting molars. That might be highly interesting to a zoologist, but apparently Cope was asking too much of the telegraph clerks. The telegram arrived totally garbled, the jaw-breaking name reduced to a strange anagram.

Almost at the same time Marsh had found a few bones of the same animal. He hastily gave the fossil a name unquestionably easier to transmit: *Dinoceras*, 'terror horn'. But this time, too, the telegraph clerk made a mistake: he transmitted to Yale 'Tinoceras', which meant nothing at all. Cope was the loser, nevertheless, for Marsh's name was readable and was accepted.

The nomenclatural defeat affected Cope deeply. For naming is an almost sacred act to biologists and palaeontologists. A man who is cheated of his priority in this respect feels almost as if sacrilege has been committed, as if the holiest principles of his science have been assailed. Cope responded by attacking Marsh with a sharpness that contrasted ill with his usually conciliatory character. He called his opponent a plagiarist, charged him with illegal appropriation of fossils, and denounced him for his constant intrigues and obstruction aimed at driving all other scientists out of the field.

As the battle continued, Cope's charges grew fiercer. Marsh, he asserted, was bribing the scouts of rival expeditions 'in order to sabotage other people's work'. He had even instructed the workers he employed to smash all the fossils which could not be shipped at once for lack of transportation, destroying them completely so that they could not fall into the hands of other collectors.

Marsh retorted, to begin with, by accusing Cope of predating his finds in order to secure priority. Then he presented a long list, which he and his assistants had compiled with great care, of Cope's scientific errors. It culminated in the statement that Cope was so monumentally ignorant that he had even managed, in reconstructing a plesiosaur he had found, to confuse the tail with the neck, so that the head was placed at the tail end.

Thanks to the faulty assembly of Cope's plesiosaur, Marsh won this round. But in the next round Cope brought up some heavier batteries. His targets were Marsh's papers on two groups of animals that had won him fame, and in which he took special pride: his monograph on toothed birds and his genealogy of the horse.

The Professorial Duel

Only a few years after the discovery of the first archaeopteryx, Marsh had dug out of the Cretaceous rocks of Kansas several fossil birds which looked by no means so primitive; in fact, many of their characteristics appeared to be strikingly modern. But like the archaeopteryx, they still had teeth in both jaws. One type, *Ichthyornis*, was presumably an excellent flyer which caught fish by power-diving into the ancient Niobrara inland sea. Another, *Hesperornis*, had only non-functioning, stubby wings, and probably lived along the coasts, diving for food.

Evolutionists burst into cheers at the discovery of these uniquely specialized toothed birds. For now it could be posited that in the Cretaceous there had probably been a large number of different descendants of the *Archaeopteryx*. The gap between modern birds and the most primitive of birds seemed to be closing, thanks to Marsh's discoveries. Huxley and Darwin both wrote enthusiastically to Marsh, congratulating him on having provided such excellent support for the theory of evolution.

On the Continent people were equally excited over Marsh's by now famous family tree of the horses. What with the toothed birds and the horses, even serious naturalists began to regard Marsh as a kind of palaeontological wizard who could conjure up whatever fossil evidence was needed to fill the gaps in a theory. Cope, of course, could not deny that Marsh had excavated the primeval horses and the toothed birds, for their skeletons were plainly exhibited in the Peabody Museum in New Haven. But Cope protested that Marsh had merely collected the specimens, and had done next to nothing in the way of scientific work on them. Marsh's horse family tree, Cope pointed out, was scarcely more than a plagiarism of Kovalevsky's work. His monograph on the toothed birds had been written for him by one of his assistants. Othniel Charles Marsh's other publications, too, were mostly written by others, Cope insinuated. Marsh was in the habit of 'buying brains' as he bought up fossils, with the same ruthlessness and contempt for humanity.

At this point Marsh lost his temper. He disclaimed all these accusations; but the tone of his replies showed how hard hit he was by these particular reproaches. Stubbornly he attempted to prove that Cope had committed serious scientific errors. He was particularly devastating about Cope's version of the 'dawn emperor'. The *Dinoceras*, as Marsh called it, had looked not at all like an elephant, but rather more like a rhinoceros. Marsh could prove that it had never had a trunk, but instead had had a number of extremely curious knobs, bumps, and bony protuberances on its head. Marsh particularly stressed its sharp canine teeth – and offered a startling theory to account for them.

The great, sabre-like tusks of the *Dinoceras* superficially resembled the terrible weapons of the sabre-tooth tigers. This fact led Marsh to believe that the animals, in contrast to all other ungulates, had been carnivores. This theory electrified the scientific community even more than had the priority dispute with Cope. For Marsh's postulate was as revolutionary as would have been the assertion that some lions feed on grass, or some cows hunt and eat rabbits. It meant that Cuvier's venerable law of correlation would have come tumbling down.

But as it turned out, Marsh's hypothesis was wrong. Like Cope, he too could read the findings wrong. Today we know that the sharp tusks of *Dinoceras* were used only for cutting water plants.

The Decline of the Dinoceras

In contrast to the titanothere, the *Dinoceras* did not remain confined to North America. The animal reached the Old World by way of the Land bridge that connected Alaska and northern Asia in the Tertiary. Osborn and Andrews, the successors of Marsh and Cope, discovered *Dinoceras* remains in the Gobi Desert; with their almost reptilian head development these Old World forms were more bizarre-looking than even the American *Dinoceras*. But in Old World and New, after a brief period of florescence, the entire family mysteriously vanished.

Once again the extinction of this group has been explained by Dollo's principle of over-specialization. Probably the animals lived in swamps, like present-day hippopotamuses, and fed chiefly on juicy, soft-leaved water plants which could crush easily and did not have to be masticated. Their teeth were not adapted to other vegetation. Hence, when the above-mentioned climatic change took place towards the end of the Early Tertiary, drying up the swamps and cutting down the numbers of water plants, the family was condemned to death on Dollo's principle that 'evolution is not reversible'.

Over-specialization is a pet word with biologists when they are not really sure just why a group of animals was fated to disappear. The Ice Age specialist Wolfgang Soergel, who has examined the life and deaths of a great many archaic animals, comments: 'The ultimate cause of the extinction of a species or group of animals is probably the fact that new conditions of life place excessively great demands on the individual. So much energy is thus absorbed for the mere preservation of life that reproductive capacity and fertility are diminished. This phenomenon is familiar to the political economist as occurring among men.'

Species that have achieved unusually high specialization have become so utterly adapted to specific conditions that, according to Soergel, any slight change in their environment causes them to be so absorbed in the struggle for life that they are scarcely able to bear young. Thus their numbers are necessarily reduced. This rule also applies to certain human races, to hunting and food-gathering peoples most of all, which nowadays confront a wholly changed world. The Bushmen, Pygmies, Eskimos, and some other tribes or peoples do not share in the population explosion which is overwhelming our planet; in spite of modern hygiene and other blessings of civilization, their numbers appear to be constantly declining.

Soergel's theory has not received sufficient attention from palaeobiologists, who are forever seeking signs of overbreeding, degeneration, environmental changes, and epidemic diseases. Perhaps it is the explanation for the extinction of large numbers of species and families, even the largest reptiles that ever lived, the great dinosaurs.

The discovery of the dinosaurs led to another bitter duel between Marsh and Cope.

The Battle of Bones

The battle over the dinosaurs, which flared up between Marsh and Cope, has been compared by Bernard Jaffe to a primordial struggle between huge beasts: the two scientists charging each other head on with lowered horns.

The Battle of Bones began in the spring of 1877. The concept of 'dinosaur' existed at the time, the word having been coined by Richard Owen to describe several European finds, such as the *Iguanodon* and the *Megalosaurus*. But it was still a purely technical term with which the public was unacquainted. Included among the dinosaurs were some skeletons, skulls, and bone fragments that Caspar Wistar, William Clark, Joseph Leidy, and Edward Drinker Cope had unearthed in the United States, including the famous duck-billed *Hadrosaurus* of Haddonfield. But on the whole the dinosaurs were so little known that Dana's book on geology, in the pre-1877 editions, did not even contain the word. No one could imagine that Marsh and Cope would discover in the course of the next two decades some eighty new species of giant saurians, among them such colossi as *Brontosaurus*, *Brachiosaurus*, and *Atlantosaurus*. Sixty to seventy feet long and weighing between thirty and fifty tons, they were the largest reptiles of all time, and after the blue whales the largest animals that have ever lived on earth.

Marsh was busy rearranging his collections in the Peabody Museum when an English schoolmaster named Lakes, who was collecting plant fossils in Colorado with an engineer friend named Beckwith, came upon a monstrous fossil vertebra more than six feet long. Lakes had never dreamed that so enormous a vertebra could possibly exist. He had heard, of course, of Marsh's researches and excavations. Therefore he quickly sketched the extraordinary find and sent the drawing to New Haven, inquiring whether Marsh was interested in the vertebra. But Marsh did not come to Colorado; apparently he thought the matter was not particularly urgent. Perhaps, too, he was passing through one of those periods in which he was more interested in social life than in the exertions of fossil hunting. At such times he would give grand dinners in his bachelor home, which was located in the museum, with a select guest list of the wealthy and the eminent. On such occasions he also liked the company of beautiful women.

Such periods of intensive enjoyment made more and more demands upon him. This may have sprung from the fact that his brusque, suspicious, and envious manner had kept him from ever having a true friend. He was said to be a very lonely man. Because of his difficult disposition he had never been able to find a wife. Or so one version went. Malicious gossips also said that his attitude towards women and fossils had been the same: a large collection was more to his liking than a single specimen.

When Marsh, for whatever reasons, did not communicate with Lakes, the schoolmaster went on collecting more such vertebrae and got in touch with Cope. He sent his specimens to Philadelphia so that Cope could see whether he wanted to buy. But Marsh's excellent system caught wind of this, and the news affected Marsh like an alarm bell. He at once ordered his agents to go to Colorado and sign a general contract with Lakes. Before Cope could even reply, the English schoolmaster had relinquished to Marsh, for a considerable sum, all the finds he had made or would subsequently make in Colorado.

According to the letter of the contract, the bones now in Cope's possession were Marsh's property; for Cope had not yet bought and paid for them. Just as Cope was on the point of delivering a lecture in Philadelphia on a 'new gigantic dinosaur', to wit, the owner of the vertebrae Lakes had sent, the schoolmaster wrote again demanding the return of the bones in the name of Marsh.

It is easy to imagine Cope's reaction. But despite his anger, he had to hand over the bones or incur a legal suit. Thus Marsh was in a position instead of Cope to describe the *Titanosaurus montanus*, a monster sixty feet long and twenty feet high, which quickly won a place in the popular mind and the colloquial language as the 'dinosaur'.

Almost on the very day that Lakes came upon the titanic vertebrae, another schoolmaster in Colorado, O. W. Lucas, discovered a no less enormous saurian bone in the vicinity of Canyon City. He, too, sent the find to Cope. And he, too, received a call from an agent of Marsh who wanted to conclude an exclusive contract with him.

Marsh's agent offered Lucas twice the price that Cope was willing to pay for future finds of the same sort. Lucas hesitated. He was a man of honour and had given Cope his word to reserve future finds for Philadelphia. In spite of the higher price, he did not want to violate his agreement. Marsh then brought up heavier guns. He denounced Cope as an unreliable fellow who broke all agreements, and urgently warned Lucas against signing any contracts with Philadelphia. Since Marsh was a university professor and the head of a great museum, while Cope was only a private collector, Lucas was at last won over. Once again Cope had been outwitted. And just at this time another misfortune descended. A severe illness prevented Cope from continuing his search for fossils for some time. His fortune, which he had invested in silver mines, was largely lost in unfortunate speculations. He was glad when the government offered him a post as consulting palaeontologist in the U.S. Geological Survey. Thereby, at least, he could continue his collecting after he was well again.

In view of all this, Marsh assumed that his greatest competitor was at last definitely out of the running. Soon afterwards he was able to celebrate further great triumphs. He discovered tremendous dinosaur graveyards. And then, with government aid, he achieved a monopolistic position that seemed to be unassailable.

In 1879 Congress decided to create a federal geological authority. It was to coordinate all research and survey work in the United States; this was the office in which Cope was employed. Marsh became chief palaeontologist of the Geological Survey. He was to place whatever fossils were found at the disposal of the Smithsonian Institution and the National Museum in Washington.

Marsh was glad to do so, since he was honorary curator in both institutions. He did not neglect his Peabody Museum, however.

He must have felt he had reached the peak of good fortune when in addition the National Academy of Sciences elected him president, in spite of Cope's protests. Major Powell, the chief of the Geological Survey, was one of his old friends. Men closely linked with Marsh and Powell held strategic posts in the federal government, in the Academy, and in the Washington museums. Thus Marsh had established a position of power such as Cuvier had once enjoyed in Paris.

He wanted to use this power to win a final victory over his old enemy, Cope, and to exult over him by obtaining Cope's treasures. He found a pretext for action in Cope's connexion with the Survey. While carrying out his duties, Cope had continued diligently to collect fossils and send them to Philadelphia. But since the Geological Survey was a government office, Cope by rights should have informed the government about his excavations and collections.

Marsh bided his time until 1889. But when Cope again tried to block his re-election as president of the Academy, he struck. Cope had collected the fossils as a government employee, he argued; therefore they should be turned over to the government. He went a step further, asserting that all of Cope's finds in Cretaceous and Tertiary strata were government property and therefore must be transferred to the National Museum in Washington.

Cope was thunderstruck on receiving an injunction from Major Powell in the name of the Department of the Interior ordering him to send his stores of bones in Pine Street to Washington. Practically, that meant that he would be turning over his entire life work to his enemy, who was Curator of the National Museum. Cope would not and could not accept that. He was prepared to fight to the last for his saurians and archaic mammals.

Scandal

From 1886 on Cope's financial situation had steadily worsened. He had parted with property and other material goods, had even been forced to sell a large collection of Tertiary mammals, of which he was particularly proud, to the New York Museum of Natural History. This museum later also acquired the Argentine fossils Cope had bought at the Paris World Exhibition. In both cases Cope had made the sales at a loss. Finally he was compelled to accept a professorship at the University of Pennsylvania, as successor to Leidy, in order to have a regular income on which to maintain his family.

The passion for collecting had made a poor man of the rich Quaker's son; he was now dependent on his salary as a professor and burdened by heavy debts. Nevertheless Cope continued to hope that he would be able to keep at least a part of his treasures, and work them up. And now the last of them were threatened by none other than Marsh!

Cope's reply to the Secretary of the Interior was blunt and undismayed. He pointed out that he had spent more than eighty thousand dollars and used up his inheritance to acquire the fossil collection in question. This collection, he mentioned, constituted a substantial portion of all palaeontological finds. All these fossils were by law and common sense his rightful property.

He then went on the warpath. For years he had been assembling material on Marsh and the activities of the Geological Survey. The moment had come, it seemed to him, to use these materials as a bombshell. For this purpose he needed the press and the interest of the public. He found an ally in William Hosea Ballou, reporter for the New York *Herald*.

Ballou studied Cope's material and wrote a series of sensational articles culminating in the charge that the Geological Survey was a nest of intrigue and corruption. Behind the mask of science gigantic speculations were being launched, government funds squandered, cliquism and nepotism fostered. Powell and Marsh were utilizing their political and scientific monopoly, Ballou wrote, like a party machine in the manner of Tammany Hall. Among other things, Marsh was turning the best jobs in the Survey, the Academy, and the museums over to the sons of wealthy and influential people, in order to elicit gifts of money from the fathers.

For two weeks the *Herald* lambasted Marsh, and Marsh answered in kind. Newspapers, scientists, and politicians passionately took sides for Cope or for Marsh. Congress investigated the accusations of both sides and concluded that Cope was the rightful owner of the fossils, but that he had gravely exaggerated in his attacks on the Geological Survey. Cope was not satisfied with this decision of the House investigating committee; he pursued the matter until the Senate, too, had to look into it.

By this time most of the Senators were thoroughly sick of the scandal. In any case they had no idea what palaeontology was good for. Consequently they arrived at a verdict that struck both scientists hard. Public moneys had been scandalously wasted on the excavation of useless old bones, they declared; all such expenditures must stop. Marsh was asked to resign. The government refused to grant any more funds for palaeontological research. For a time it looked as if the entire science of palaeontology would be discredited in the United States, and that its ill-repute would linger on for years. Cope died in 1897 in the midst of his hecatombs of bones. Marsh, too, had meanwhile exhausted his means and for the first time in his life had to accept a salary from Yale. He died two years after his rival. But his successor in the Geological Survey was not one of his own assistants; it was Cope's friend and onetime disciple, Henry Fairfield Osborn.

Thus Cope won from beyond the grave. For Osborn also took over the direction of the American Museum of Natural History in New York, which had purchased the greater part of Cope's collections. Through Osborn and his staff of scientists, the New York museum became the home of the leading palaeontological collection in the world. Osborn, hailed in learned circles as the 'Darwin and Huxley of the New World', was able to recover lost ground for his science and make its fascination felt to the general public.

From Dragon to Giant Lizard

The dinosaurs that Marsh and Cope had discovered have become the most popular and the most frequently described of primordial animals. To this day the average man is sure that all dinosaurs were gigantic monsters; to this day popular science repeatedly poses the riddle of what global catastrophes must have destroyed these colossal reptiles. For to this day Cuvier's cataclysms haunt the minds of the multitude.

The unfortunate dinosaurs have thus been seriously mistreated by *Homo sapiens*. They have been impressed into fantastic fiction, from Conan Doyle's *The Lost World* to the monster and horror films of our day. They figure in innumerable cartoons as clumsy, stupid monsters being hunted by primitive man. They are misrepresented occasionally even in serious writings: thus in 1924 the German palaeontologist Edgar Dacqué, curator of the Munich state collections, published a work entitled *The Ancient World, Legend and Mankind*, wherein he asserted that the legends of dragons, which are found among so many peoples, can be ascribed to a kind of racial memory of primordial times. Man's Mesozoic ancestors, Dacqué theorized, must have lived with dinosaurs and other great reptiles; and these beasts would have made so strong an impression upon their developing minds that the impression became hereditary and has continued on in our imaginations.

This idea is intriguing but it stands on rather weak foundations. For the Mesozoic ancestors of man, that is, the contemporaries of the saurians, are – as was known even in Darwin's day – small, reptile-like mammals about the size of shrews. These primeval mammals scarcely came into contact with the giant saurians, which were mostly harmless vegetarians anyhow – unless, that is, a hungry Jurassic or Cretaceous mammal occasionally ate a saurian egg. It is hard to imagine that these primitive mammalian ancestors could have handed any racial memories of the dinosaurs down to us.

The occasional finding of a fossil saurian may here and there have led to the birth of a dragon legend. Some scientists trace our ideas of dragons back to primitive man's encounter with giant snakes, crocodiles, and other large reptiles. Others suggest that the cave bear gave rise to the legend. Nevertheless, the dragon of fairy tales bears no relation to giant snakes, crocodiles, cave bears, or other living or dead creatures; on the contrary, the dragon of our imagination does indeed resemble a giant saurian. We do not know why that is so, for it is incontestable that man never saw the giant saurians.

Counter to the usual view, the dinosaurs were by no means all huge. Many species were no larger than present-day reptiles. Nor must they be thought of as clumsy, mentally laggard monsters whose existence was nipped short because they lacked brains. In reality the primordial dragons did not disappear abruptly at the end of the Mesozoic. In the course of four great periods, Permian, Triassic, Jurassic, and Cretaceous, they gradually increased in dominance and then gradually diminished until they disappeared like so many other groups of animals whose clock had run down for one reason or another.

Moreover, the saurians by no means vanished without leaving any descendants. On the contrary, they are represented today by a highly flourishing posterity. All higher vertebrates are descended directly or indirectly from saurians. Strictly speaking, moreover, tortoises, crocodiles, and *Rhynchocephalia* should still be reckoned among the saurians, for they existed in the time of the saurians and have scarcely altered since. Snakes have sprung from the same root as the great mosasaurs, the 'sea serpents' of the Cretaceous period, to whose discovery and study Cope so significantly contributed. Lizards, too, are modest descendants of the same great ancestors.

Mammal-like saurians which ran on all fours and were probably largely carnivorous existed in very great numbers during the Permian period. Such pelycosaurs, therapsids, and ictidosaurs have been found almost everywhere on the globe, particularly in South Africa, and continued on into the Triassic. Two groups,

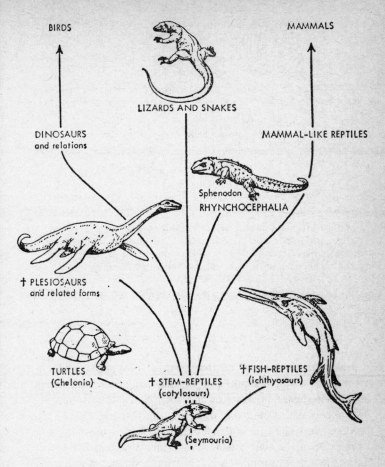

BIRDS

MAMMALS

LIZARDS AND SNAKES

DINOSAURS
and relations

MAMMAL-LIKE REPTILES

Sphenodon
RHYNCHOCEPHALIA

† PLESIOSAURS
and related forms

TURTLES
(Chelonia)

† STEM-REPTILES
(cotylosaurs)

† FISH-REPTILES
(ichthyosaurs)

(Seymouria)

Family tree of the reptiles, showing that the saurians have living
descendants.

the theriodonts and the ictidosaurs, pass so gradually into the
first primeval mammals that with some Triassic finds there is no
saying whether they should be classed among the reptiles or the
mammals.

Thus the gigantic types of the dinosaurs, which in the popular
mind are the true saurians, constituted in fact a minority. Their
appearance and life-pattern excited intense general interest from

Mammal-like saurians. The dimetrodon, a dangerous predator, surprisingly resembled the edaphosaur; both had spiny processes growing out of the backbone, which supported a sail-like flap of skin. The purpose of this curious formation is unknown.

the time of their discovery. And a great deal was found out about them as early as the days of Marsh and Cope.

> '... No problem bothered him a bit;
> He made both head and tail of it.'

In the summer of 1877 the first great find took place. The railroad men of Fort Laramie, Wyoming, had often come upon monstrous bones in quarries and rock clefts, 'fearful-looking fabulous beasts, gigantic nightmare figures that frightened men, women, and children and caused even horses to shy', as a chronicler of the time related. Now the railroaders heard that there was a professor in New Haven who threw money around to buy up such atrocious bones.

The railroad workers sent Marsh a letter inquiring whether he would be willing to buy fossils from them on a steady basis and employ them in excavation work. The letter closed on the frankest of notes: 'We have been told you are a keen geologist and a very rich man. We would like to meet you, especially for the second reason.'

In this wise, the dollar-hungry railroaders of Fort Laramie led Marsh to one of the largest dinosaur graveyards that has ever been located. Everywhere in the United States people were soon talking about the giant saurians that Marsh was bringing out of the 'Bone Quarries'. There was, for example, the skeleton of a brontosaur, a peaceful herbivore some sixty feet long. Marsh paid thirty thousand dollars simply to have it mounted in Yale University. Even stranger was the stegosaur, a heavily armoured dinosaur that seemed like a baby beside the brontosaur – it was only seventeen feet in length – but whose form was quite remarkable. On its back it bore huge triangular bony plates; at the tail end it had four powerful stings with which it must have administered rebuffs to any attacker.

The stegosaur was the first dinosaur in which a kind of second brain was discovered in the hip region, a brain, in fact, which was far larger than the brain in the head. Marsh thought the 'hip brain' must have served chiefly to direct the movements of the mighty tail. But it is possible that the rear nervous centre also assumed other functions of the brain. Such 'posterior brains' were subsequently found in other types of saurian. The curious organ inevitably inspired a good many humorous observations and verse. The stegosaur has been proposed to philosophers as a model since, as Bert L. Taylor wittily put it:

> Thus he could reason *a priori*
> As well as *a posteriori*. . . .
> If something slipped his forward mind
> 'Twas rescued by the one behind.

More than two hundred and fifty genera of dinosaurs are now known. A considerable number of them were indeed giants. The strata on the borderline between the Jurassic and Cretaceous in the United States turned out to be crammed full of saurian bones. From Alberta Province in Canada all the way down to New Mexico and Texas palaeontologists have come upon extensive saurian graveyards.

Brontosaurus and *Brachiosaurus* are today held to have been the largest saurians that ever lived. For a while this honour was also conceded to a close relative of the two, *Diplodocus carnegiei*, a Wyoming saurian named after multimillionaire Andrew Carnegie. Carnegie had followed Peabody's example and given large sums of money for palaeontological research. He founded the Carnegie Museum in Pittsburgh and equipped formidable expeditions which found so many dinosaurs in Colorado, Wyoming, and especially in eastern Utah that the sites were called saurian quarries. One of these vast areas in northeastern Utah and northwestern Colorado was set aside as a nature preserve in 1915, and named Dinosaur National Monument.

On one of these expeditions Henry Fairfield Osborn and his assistant Walter Granger noticed some dark-brown ledges cropping out of the flowers, grasses, and shrubs of the prairie. On nearer view, the two scientists realized that the rocks were in reality pieces of stupendous, weather-beaten saurian bones. In the immediate vicinity an exceedingly thick thighbone protruded some six feet out of the ground. Osborn and Granger tried to pry out the bone. But it continued on and on; more bones came to light; and finally the greater part of a *Diplodocus* skeleton emerged.

This huge saurian had an unusually long neck. In addition, its nostrils were not at the front of the snout, but far back on the head. This is a peculiarity of whales, sea cows, marine saurians, and other aquatic creatures. It led to the conclusion that at least some types of giant saurians must have lived in swamps. Since they possibly could not swim, they may well have lumbered through the mud at the bottom of the swamp, grazing on water plants and from time to time stretching their long necks and heads out of the water to breathe.

A Fighting Machine

Even weirder in appearance were the *Ceratopsids*, the horned 'rhino saurians' with their armour plate, combs, and spines, which were discovered in the Cretaceous layers of the American West and later also in Mongolia. In 1887 Marsh found a very odd-looking horn in the vicinity of Denver, Colorado. At first he thought it the horn of some gigantic fossil buffalo. But further excavations in Colorado and Wyoming proved that it must be a reptilian horn. In 1889 Marsh's collectors brought to light the first skull and soon afterwards the entire skeleton of the horned saurian. This animal, which was called *Triceratops horridus*, was perhaps the most amazing of all the saurians.

The colossus was approximately twenty-five feet long, stocky in build, with armoured skin and a rhinoceros-like horn on its nose. Its most curious feature was the monstrously enlarged, bony neckplate, which was in addition studded with large sharp spines. Above the eyes the creature had two additional horns. This probably peaceful vegetarian had converted itself into a living fortress which must have been invulnerable to the attacks of even the fiercest predators.

But there were predatory saurians in the Upper Cretaceous which might even have managed to deal with the horns and armour plate of *Triceratops*. The finest skeletons of this great carnivorous dinosaur were dug up at Hell Creek in northern Montana by Osborn's friend Barnum Brown. It had a length of more than forty-five feet, walked on two legs, and had a huge head with a truly terrifying row of knife-edged teeth. Osborn gave this largest of the predatory dinosaurs the imposing name of *Tyrannosaurus rex*. Many imaginative souls insisted that *this* saurian must have been the actual prototype of the legendary dragon. William Diller Matthew, a colleague of Osborn's at the Museum of Natural History in New York, concluded that the tyrannosaur must have possessed such fearful weapons because its chief prey was the heavily armoured triceratops. As in modern naval warfare the armour plate of ships constantly increased in thickness and the armour-piercing power of torpedoes and shells similarly increased in efficacy, so in the primordial struggle for existence the armour of the triceratops and the dental weapons of the tyrannosaur developed to extremes.

But it is not altogether certain that the tyrannosaur was really a predator that attacked and tore to pieces other giant saurians. Some palaeontologists argue that the big carnivorous dinosaurs were far too clumsy to go hunting, and probably subsisted on carrion.

The skeleton of the tyrannosaur was dispatched to New York and exhibited, along with many other finds, in its Museum of Natural History. There the indefatigable Osborn trained a team of highly qualified palaeontologists and fossil hunters. The era of systematic fossil hunting began. If an area seemed promising, its geological aspects were first thoroughly studied; maps were made and a plan of attack prepared before the dig began. Among Osborn's most successful associates was the same Barnum Brown who had discovered the tyrannosaur, 'the most terrible fighting machine that nature ever conceived'. Barnum Brown has gone down in the history of science not only as a dinosaur collector, but also as an expert on the Tertiary period in the Americas and Asia.

Married to Dinosaurs

Seldom has a palaeontologist had so genial a destiny as Barnum Brown. The era of slanderous and bitter 'battles of bones' was over at last. Palaeontologists were learning to appreciate and esteem one another. And Brown was an unusually likeable chap, as well as the subject of a host of good-natured anecdotes.

Like Marsh and Cope, Brown was a farmer's son. He had studied geology at the University of Kansas, then joined the team of Henry Fairfield Osborn and quickly demonstrated his remarkable flair for finding fossils. At the museum, under Osborn's guidance, he became a brilliant reconstructor of primordial reptiles. The skeletons as prepared by Brown, and the pictures of the animals in life that Charles Knight painted on the basis of these reconstructions, were soon regarded as exemplary.

He was ably assisted by his wife Lilian, known as Pixie among colleagues. The two had met on an expedition to India and Brown hurried his bride straight from the altar to a wedding trip into the jungle. The object of this trip was not Brown's beloved dinosaurs, but fossil mammals in the Sivalik Mountains of northern India. The Sivalik fauna (elephants, primitive giraffes, antelopes, and protoanthropoids, along with sabre-toothed tigers and protocanines the size of bears, and those strange *Sivatheria* which are regarded as the characteristic animals of the Sivalik strata) proved to be a highly illuminating counterpart to the Pikermi fauna of ancient Greece. The *Sivatherium*, named after the Sivalik Hills, was a short-necked giraffe about the size of an elephant whose head was crowned by elk-like shovel antlers – one of the most fantastic beasts of the Tertiary period.

Lilian Brown took part in her husband's hunt for another and even stranger giant animal. This was the *Baluchitherium*, probably the largest land mammal that ever lived. Remains of this

313

The *Baluchitherium*, a hornless rhinoceros of Oligocene Asia. Seventeen feet high, it was larger than any other land mammal. Here Osborn shows it matched against a present-day Indian rhinoceros.

mighty hornless rhinoceros have been found in Baluchistan and Turkestan. Along with Roy Chapman Andrews, the Browns traversed some of the most inhospitable regions on earth in their search for more bones of the hornless rhino. On 5 August 1922, on the way across Mongolia, Andrews dug up a number of bones in the Tsagan-Nor basin which, back in New York, were later assembled into a skull. So enormous was this skull that Osborn at first thought his associates must have made a mistake and put together fragments of several skulls. But later finds proved that the baluchitherium had actually stood seventeen feet high and been more than twenty-two feet long. There were other kinds of enormous species of rhinoceros in Asia, some hornless, some with their horns not on their noses but growing from the middle of the forehead. One of them, the *Elasmotherium*, probably survived deep into the Ice Age and was possibly a contemporary of man.

The Hell of La Brea

What Barnum Brown found in the Sivalik Hills might foster the impression that here a whole earthly paradise had flourished, and then, with one fell swoop, been extirpated. This was not the only place where nature seemed to have inflicted inconceivably cruel mass destruction upon its creatures. Other palaeontologists, among them John C. Merriam and Chester Stock, had come upon similar hecatombs of fossils in lava pits and asphalt lakes in the western United States. The asphalt marsh of La Brea, discovered by John C. Merriam and located in what is now the environs of Los Angeles, achieved a macabre fame among such burial places.

For many years a whole staff of palaeontologists plumbed the vicious, sticky asphalt. The number of Tertiary animals drowned in this swamp in the course of time is estimated at several millions. Mastodons, brontotheres, giant sloths, camels, and horses were swallowed up by the treacherous pitch. Their death struggles attracted vultures, wolves, and sabre-toothed tigers which were in their turn caught in the tar and buried together with the prey they had anticipated.

Streams from the Californian mountains probably flowed into the La Brea area. The pools of water forming on the surface of the asphalt lake lured thirsty animals from the surrounding desert. Possibly the asphalt was also repeatedly covered by volcanic ash so that animals that unsuspectingly approached could not recognize the terrible death trap. Similar mass graves have been found in Siberia, China, and the Gobi Desert. The fact that so many animals died in so confined an area indicates that the asphalt marshes attracted fresh victims for thousands upon thousands of years.

But cruel though such tricks of nature are, the finds in the asphalt lakes and other such mass burial places are peculiarly useful to palaeontology. The countless individuals that died a natural death in most cases turned to dust and left no trace of their mortal existence. The victims of catastrophic events, on the other hand, remained preserved in the rubble of ravines, under blankets of volcanic ashes, or preserved in the sterile syrup of asphalt. The majority of discovered fossils of terrestrial vertebrates have been drawn from such mass graves.

Yet we must not draw false conclusions from the presence of such sites. In our human history mass annihilation by earthquakes, volcanoes, floods, conflagrations, and other catastrophes, for all the terror and suffering they caused, have scarcely played a decisive part in the outcome of things. Similarly, in the remote past of this planet, dramatic mass deaths have been only an exception, inconspicuous, natural death the rule.

The Darwin and Huxley of the New World

For decades Osborn and his able associates combed all the regions of the world for fossils. The 'Darwin and Huxley of the New World' could draw on a wealth of material such as has been available to scarcely any other palaeontologist. Osborn used it to coordinate a variety of theories about the evolution of species. Although his teacher had been Cope, the Neo-Lamarckian, Osborn belonged to no scientific school and refused to take a partisan position on biological questions. Instead, he incorporated the ideas of Buffon, Lamarck, and Darwin, as well as the laws of heredity and the experiments of modern geneticists, into a new comprehensive theory.

Buffon, he pointed out, attempted to explain evolution by environmental influences, Lamarck by the use or non-use of organs, Darwin by natural selection and mutational genetics. But none of these hypotheses sufficed to explain in full the mysteries of evolution. Only the four processes together provided the basis for the evolution of organisms.

Osborn formulated what he called his 'law of the four inseparable factors of evolution', which ran as follows: Life and the evolution of organisms are constantly guided and influenced by four processes: by heredity, the development of the individual, by environmental influences, and the effects of selection. These four natural processes have been inseparably linked from the beginning and are constantly interacting with one another. As soon as one of these factors stimulates or initiates a modification, the other three factors react, with the result that the modification becomes a fundamental change.

In thus coordinating the various theories of evolution, Osborn purged much of the doctrinaire spirit from the science. His work inaugurated a new era in which bold theories and dramatic excavations went out of style, giving way to closely detailed researches. Step by step, with the aid of modern physical and chemical methods of investigation, the ages of geological formations and their organisms were determined. The history of the earth, as the scientists wrote it, changed more and more from a thrilling, highly imaginative novel to a sober but no less fascinating work of non-fiction.

Osborn took a great part in this development. When he died in 1935, most American palaeontologists, whether they were Neo-Darwinists, Neo-Lamarckians, or adherents of other schools, concurred on one point: they could rightly consider themselves 'Osbornists'.

Book Six

A Thousand Years are as a Day

Millions and whole mountain ranges of millions of centuries will pass,
within which forever new worlds will be formed.

IMMANUEL KANT, 1775

The Earth's Clocks

'Various reasons cause us to believe that the new radioactive substance contains a new element to which we wish to assign the name of radium.' So went the note read to the Academy of Sciences in Paris on 26 December 1898, by the physicist couple Marie and Pierre Curie. 'The radioactivity of radium must be tremendous.'

At the time no one could imagine what weight this discovery of radioactivity was to have – not only for modern atomic physics, but for the whole history of the human race. For the time being, it did not occur to anyone to apply the transformation of radioactive minerals to measurements of geological time. Decades passed before the perception came that the 'radium clock' constituted the best and most reliable method of determining the age of the earth and the duration of various geological ages.

Ever since geologists and palaeontologists had freed themselves from Biblical chronology, they had handled figures and dates in an expansive manner. They boldly began reckoning in millions of years; they flung down seven, eight, nine, and ten place numbers as though to underline the 'millions of centuries' of which Kant had spoken. But these enormous figures seemed also to have introduced an element of uncertainty and flippancy into the sciences of the earth's past. The dates given by various experts differed enormously. All kinds of methods of dating were developed, but none of the figures was susceptible to exact proof. The various estimates were all so questionable and so contradictory that the scientific voyages into the earth's remote past bore an amazing resemblance, at least in the eyes of the general public, to dreams and visions.

Since the days of Buffon's historic experiment with the two metal spheres, numerous physicists, geologists, biologists, oceanographers, and astronomers had attempted, with the aid of earthly or celestial factors, to determine the age of our planet. During the period of Lyell and Darwin general acceptance was given to a method devised by the great English physicist Sir William Thomson, later Lord Kelvin. Employing Buffon's principle, Thomson tried to estimate, on the basis of physical observations and logical deductions, the time the earth would have needed to cool from a glowing hot state to the heavily crusted, inhabitable planet it is today. To the great disappointment of the evolutionists, however, Thomson arrived at a minimum figure of twenty million years and a maximum of four hundred million. The range was too great, the figure too imprecise, although Lord Kelvin had been one of the first scientists to venture to think in terms of millions of years. Many of Darwin's followers preferred different and apparently more practical methods of reckoning which yielded more exact results than Lord Kelvin's.

One of these methods was based on the fact that in the course of time thick layers of sediment have formed on the floor of the seas. A large portion of our rocks are derived from such sediments. The geologists tried to measure the thickness of various layers of sediment and to calculate the time needed to deposit them. That, too, was an extremely uncertain enterprise; for to this day nobody knows just how thick the thickest layers of sedimentary rock really are. Nevertheless, by this method a maximum age for the earth of approximately one hundred million years was reached.

The figure seemed to correspond with a method of calculation proposed by the English astronomer Edmund Halley in the eighteenth century and further developed by the French physicist Nicolas Joly. Joly tried to determine how much salt is annually leached out of the rocks and washed into the ocean. The entire salt content of the oceans, he concluded, would then indicate how long the seas have existed on our planet, thus providing the preconditions for life on earth. This method produced figures between eighty and one hundred million years.

For the evolutionists that was still too short a time. Nevertheless most scientists around the beginning of the twentieth century had to accept the view that the fixed crust of the earth was about one hundred million years old. Men like Ernst Haeckel fully realized that this figure stood in outright contradiction to the periods of time required by evolution. What was more, the geological methods of calculation did not agree with the observations and conclusions of the astronomers.

The astronomers were far more generous and bold than the geologists. 'All astronomical data,' Fritz Kahn summarized their conclusions, 'the statics of the universe, the mechanics of the Galaxy, the movement of the stars in star clusters, the dynamics of the solar system, the orbit of the moon and the behaviour of the earth in her orbit, indicate that our planet is between three and seven billion years old.' Thus, long before the present theories of the universe were developed and long before the results of astrophysical research were available, astronomers had arrived at figures in billions.

But what good was this tremendous leap backwards in time if the earth itself offered no evidence for such a great age? The problem of the age of the earth became so difficult and confusing that in 1929 a special conference of noted scientists was called in the United States to search for the best method of dating. Professor Adolph Knopf of Yale acted as chairman. He soon hit on the idea of using atomic disintegration as a measurement of geological time.

The Uranium Clock

In the meantime physicists had intensively studied Marie Curie's great discovery, radioactivity. The radioactive element uranium, they had found, did not change into radium, but continued to break up. Radium is only one of the fourteen stages of uranium disintegration. Most natural radioactive substances are derived by a series of transformations from Uranium 238, Uranium 235, and thorium. The last and ultimately stable product of the 'radioactive series' is an isotope of lead which differs from ordinary lead in atomic weight.

Knopf and his associates proceeded from the following considerations. In various places in the earth's crust uranium-bearing minerals are found. The times necessary for the various stages of transformation are known. Every uranium-bearing mineral has a certain content of lead and helium. If the percentage of these disintegration products is compared with the quantity of still untransformed uranium in the rock, the duration of the process of transformation and thus the age of the rock can be calculated.

In 1941 Knopf published his book *The Age of the Earth* which discussed in detail the possibility of using this radioactive process to determine geological ages. The principle he applied has been described, with some simplification, by Fritz Kahn:

'How long uranium has existed in the world, we do not know. But we can say how long it has lain on earth in the particular spot where we find it today. One gramme of uranium provides by its disintegration .000,000,0076 grammes of uranium lead annually. From the formula:

$$\frac{\text{Quantity of uranium lead}}{\text{Quantity of uranium}} \times 7,600,000,000$$

we obtain the number of years that have passed since the uranium was deposited at its present site.'

Let us say, for example, that certain quantities of uranium are found in rock formed in the Carboniferous (Mississippian and Pennsylvanian) period. With the aid of the formula the time necessary for the formation of the lead and thus the age of the rock can be reckoned. Uranium-bearing pitchblende, however, is a mineral from the deep strata, which like granite was contained in the magma, the molten interior of the earth. In every geological period such fluid masses have been brought up to the surface where they cooled and crystallized into rocks and ores. Hence the uranium-bearing minerals must have intruded into the rest of the Carboniferous rocks, and are therefore somewhat younger than the surrounding rocks. The calculation therefore yields a minimum value; the Carboniferous formation must be somewhat older.

The method can be applied to all ages of the earth, for pitchblende is found in almost all formations. Consequently, radioactive minerals were collected from everywhere on the globe, were analysed for uranium and lead content. It had to be determined, in order to eliminate sources of error, whether the pitchblende really belonged to the formation in which it was found, whether it had lain there unchanged since it was deposited, and whether it had been influenced by air and weathering. The quantity and type of lead also had to be precisely determined, for it might be thorium lead, uranium lead, or ordinary lead that chanced to be in the rock. Finally the quantities of helium had to be determined, for although helium, being gaseous and easily lost, is generally less reliable an indicator of the amount of radioactive disintegration than solid lead, it can serve to check the other figures.

A great many painstaking researches, a great many analyses and corrections were therefore needed before the British scientist Frederick E. Zeuner could publish his book *Dating the Past* in 1946. Zeuner described not only the uranium clock, but other geological clocks that permitted a variety of checks and rechecks.

The Fluorine Test

Are there any characteristics by which a fossil bone reveals its age? The palaeontologists have naturally posed this question repeatedly. As early as the mid-nineteenth century it occurred to French and English chemists to look to the element fluorine for determinations of age. Traces of fluorine are found in many fossils; in a calculable period of time every bone that lies in a fluorine-containing strata of earth will slowly and steadily increase in fluorine content. But since the quantities of fluorine in the earth and in ground water differ, the first fluorine datings were highly dubious, so that the matter was finally abandoned. The method was completely forgotten.

It was revived once again during the Second World War. A young British anthropologist named Kenneth Page Oakley had been assigned to look into the question of whether water containing fluorine can heal or prevent dental caries. Since Oakley happened also to be a passionate geologist, fossil investigator, and prehistorian, he developed in the course of his studies, as a kind of by-product of his concern with dental decay, the fluorine test. After the war Oakley had numerous opportunities to try out the test. Fluorine dating permitted a relative though not absolute determination of a fossil's age. First the fluorine content of the earth or rock strata is tested, and if possible several comparison fossils from the same strata; tests are then made for the percentage of fluorine in the find in question. With this data it is possible to calculate how long the bone has lain in the earth.

The fluorine test became world-famous in 1953 when it was used upon a dubious human fossil, the Piltdown Man with its apelike jaw and highly developed brainpan, which would not fit into accepted views of the descent of man. The fluorine test exposed the Piltdown fossil as a clever forgery. Oakley and his associates were able to test a number of other human fossils by the same method. The test confirmed the accepted dating of some prehuman and early human fossils, corrected others. Many animal fossils have since been subjected to the fluorine test and other microchemical tests, which include examination of the iron content, for example. Thus fossils have developed clocks of their own in the form of chemicals absorbed by them – though unlike the absolute uranium clock (and certain others, such as the potassium-argon method) these chemical clocks are relative.

Radiocarbon Measurement

If the uranium method may be considered the hour hand and the fluorine method the minute hand of the geological clock, the radiocarbon test might be called the second hand. Like uranium measurement, it is based on radioactivity. It has been found that radioactive processes can be used for dating not only the remotest geological ages, but also organic remains from the recent past.

Willard Libby, the American chemist who discovered the radiocarbon method, is one of the atomic scientists who was involved during the last war with nuclear fission and the development of the atomic bomb. In the course of his researches Libby found that throughout organic nature radioactive carbon, C-14, is present alongside normal carbon, C-12, although in inconceivably small quantities. All the carbon dioxide in the atmosphere can be considered slightly radioactive. The radioactive carbon is generated in the upper strata of the atmosphere under the influence of cosmic radiation. All plants on our earth assimilate radioactive carbon dioxide in the course of their metabolism. By eating plants, animals and men constantly take C-14 into their bodies. As long as the organism is alive, the loss of C-14 by atomic disintegration is balanced by absorption of C-14 in food. The radioactive clock begins to run only after the death of an animal or plant. Then, slowly and steadily, as precisely as a clock, the C-14 disintegrates and is transformed into normal carbon, C-12.

In 1950, after innumerable experiments, Libby learned how this phenomenon could be used for measuring time. An organism that contained at the time of its death a gramme of C-14 (this is only for purposes of illustration; usually the actual content of C-14 is very much smaller) will have only half as much C-14 5560 years after its death. After the passage of 11,120 years only half of the half will be left, after 22,240 years only a sixteenth, after 44,480 years only 1/256, and so on. The method was simple; the real problem was to construct physical apparatus to detect with precision the most minute traces of C-14.

At first it seemed as if the radiocarbon method could be used only for dating organic remains from historical times. Libby analysed wood from the tombs of Egyptian kings, Indian arrowheads, a piece of linen in which one of the much-discussed Dead Sea Scrolls was wrapped, and other archaeological matter. But then he and his associates succeeded in pushing their dating back to twenty-five thousand years and determining the ages of the European cave-paintings, Pleistocene giant sloth bones, and petrified trees of the Wisconsin Ice Age. By 1960 the apparatus had been so refined that animal and plant remains more than sixty thousand years old could be dated.

The measuring implements have since been still further improved, and refinements have been introduced into the uranium and fluorine methods as well. We can already say that these 'geological clocks' are going very accurately, much more accurately than all previous time-measurements. The optimistic 1952 prophecy of Frederick Johnson, an associate of Libby, is already being confirmed: that the means at last exist for creating a universal scale of chronology.

How Old is the Earth Really?

Nevertheless, the most important of the new methods of dating remains the uranium clock. For by its use geologists and palaeontologists could obtain more accurate figures. The various ages of the earth, beginning with the Cambrian period, could be dated with fair exactitude from the uranium-bearing rocks in the formations. There are no longer any significant disagreements about the duration of the twelve epochs within the five hundred million years of the Phanerozoic aeon. But what lay before the Cambrian period? How long had the Precambrian, the so-called Cryptozoic aeon, lasted? In other words, how old is the earth, really? In the Cambrian seas of five hundred million years ago all the stocks of invertebrate animals already existed. From that it has been concluded, as the Swiss palaeontologist Bernhard Peyer put it, 'that the earth must have been inhabitable before the Cambrian for a period about twice as long as from the Cambrian to the present'.

Thus life has existed on our earth for about one and a half billion years. That was the approximate age of the oldest rocks dated by the new uranium method soon after its discovery. If another half billion years were added during which the earth was in a gaseous or molten condition, two billion years was arrived at as the age of the earth. But refinement of the uranium clock soon invalidated this reckoning. In North America and South Africa minerals were found of an age between 2·3 and 2·6 billion years. These were minerals of rocks from the magma, which had intruded into a framework of older rocks. Hence the two and a half billion years must represent a minimum. In Transvaal, shortly afterwards, the mineral uranitite was discovered and dated at almost 2·8 billion years. A piece of rock on the Kola peninsula has recently been dated at 3·4 billion years. Some geologists therefore estimate the age of the earth at between three and a half and four billion years. The human imagination boggles at such figures.

Precambrian: Let there be Light

In Archaic times western North America, together with Greenland, formed one continental mass which geologists call Laurentia. The gneiss strata and other rocks of Laurentia are among the oldest on earth; minerals found in them have been dated at 2·55 to 2·64 billion years.

Around the middle of the nineteenth century, long before the uranium clock had been discovered, it was nevertheless known that the deepest strata of gneiss near Ottawa, Canada, must be extremely old, Palaeontologists of the time were therefore amazed when the Canadian geologist William Edmund Logan reported that one of his associates had discovered the remains of organisms in such rocks. He had come upon bulblike formations of limestone within the gneiss strata; some of the bulbs were the size of a human head. In spite of their size these bulbs resembled groups of unicellular animals which live in pretty and frequently very complex limestone structures – the foraminifera.

Foraminifera have contributed more to the building of the earth than all other organisms put together. They already occupied the Palaeozoic seas, and they exist today in unimaginable quantities. Some fifty million square miles of sea floor, one fourth of the surface of the earth, is covered by a sediment formed by a single type of foraminifera, *Globigerina bulloides*. Some modern measurements indicate that there are sedimentary strata of foraminifera in the ocean abysses more than eleven thousand feet thick. Towards the end of the last century a German zoologist, Max Schultze, took the trouble to calculate the number of foraminifera shells in an ounce of sand from the Italian coast. He found one and a half million; approximately half the sand consisted of the chalky shells of these one-celled builders.

Even in the Paleozoic seas, the shells of foraminifera accumu-

329

lated in layers of enormous thickness. In the Mesozoic, foraminifera contributed to those strata which later became the chalk cliffs of England and the vast limestone areas of North America. For five thousand years foraminifera have also visibly been part of human civilization. The Egyptian pyramids, Hagia Sophia, the great monuments of Paris, and countless other buildings were constructed out of foraminiferous rock.

Some types of foraminifera, especially those of the Tertiary period, attained extraordinary size. While one-celled organisms are usually microscopic, these gigantic protozoa reached the size of a nickel. The zoologists call them *Nummulites*, 'coin animalcules'. About the beginning of the Christian era the Greco-Roman historian Strabo discovered vast numbers of such nummulite shells in the stone blocks of the pyramids. He wondered about their shape, and came to the curious conclusion that they must be the petrified noon meals of ancient Egyptian masons. The pyramid builders must have eaten lentils, he reasoned; and the remains of their lentil porridge had stuck to the blocks and gradually turned to stone.

Dawn in Laurentia

At the time Logan identified the bulbous lumps in the Ottawa gneiss as foraminifera shells, little was known about unicellular animals. Many foraminifera have spiral shells; consequently they were compared with ammonites and regarded as tiny fossil relations of squids. For a while no one raised the question of how *Eozoon canadense*, 'the Canadian dawn animal', as Logan's find was called, could possibly grow as large as a fist, a cabbage head, or a human head.

In the following decades the 'dawn animals' were studied by some of the foremost specialists in fossil micro-organisms. If one of the lumps were slit open, the cross section looked like a compressed, many-branched shrub. It surprisingly resembled the interior of a gigantic foraminifera; for the chambers of any single foraminifer, which are strung together like a string of beads, also look like the twisting branches of a shrub. Authorities such as the famous oceanographer William Benjamin Carpenter, the English palaeontologist William Dawson, the German geologist

Georg von Gumbel, and the protozoon specialist Max Schultze supported Logan's view that the animal was the 'oldest remains of life on earth'. Only a few persons energetically contested the organic origins of the Eozoon.

Gradually, however, it became clear that the foraminifera are unusually highly developed unicellular animals. Life could scarcely have begun with them. In all probability the oldest living organisms had been calcareous algae and other one-celled plants, delicate specks of protoplasm which left no trace behind in the rocks. Only when the unicellular animals began breaking up the silicic acid compounds in the seas and making fine-grained skeletons and shells could their remains be preserved. These graceful creations, 'nature's art forms', as Ernst Haeckel once called them, first appeared in late Precambrian times. By then an abundance of multicellular animals existed. Hence, more and more scientists began expressing doubts about the organic nature of the Eozoon.

Among the last defenders of the 'dawn animal' was none other than Darwin. He went so far as to describe this supposedly oldest organism: 'The Eozoon belongs to the most lowly organized of all classes of animals, but is highly organized for its class; it existed in countless numbers, and . . . certainly preyed on other minute organisms, which must have lived in great numbers.'

Soon after these words were written *Eozoon canadense* had to be stricken from the list of fossil organisms. Similar formations had meanwhile been found in Europe. A German oceanographer, Karl Möbius, was able to demonstrate conclusively in 1878 that the mysterious lumps were not fossils, but inorganic mixtures of minerals. Darwin admitted that he had been mistaken, and that the question of Precambrian fossils remained unsolved.

Graphite and Petroleum

The controversy over the Eozoon spurred the search for traces of the oldest Precambrian organisms. Graphite embedded in ancient rock soon came under discussion. Since graphite is a variety of carbon, it was assumed that like coal it arose from the gradual carbonization of vegetable remains. Thus towards the end of the last century the idea prevailed that the occurrence of graphite in

Precambrian formations was evidence for some of the oldest organisms, whose forms had entirely disappeared. The graphite, it was held, might represent the carbonized remains of archaic algae.

But graphite is also formed inorganically, by the heating of iron carbides; moreover, meteorites which fall to earth from space frequently contain graphite. Thus the graphite embedded in the most ancient rocks is no sure proof of any life.

For a while petroleum was considered the decay product of myriads of archaic microorganisms. One theory of the origin of petroleum – which is by no means universally accepted – holds that dying plankton which is quickly covered by mud or lime deposits in salt-rich and oxygen-poor lagoons does not decay, but passes under the influence of bacteria and catalysts into a characteristic mixture of hydrocarbons. Thus, it is argued, oil slates, tar, asphalt, petroleum, and methane were formed out of the onetime microflora and microfauna of the seas.

Petroleum, because of its light weight, constantly tends to move into regions of lower pressure. This tendency is what brings it to the surface – to the delight of oil companies. But it also complicates the problem for geologists who are seeking to determine when and where the petroleum formed. The present sites, the rocks in which oil is stored, are by no means identical with the rocks in which it presumably originated. For that reason it has been impossible to determine definitely what organisms have produced petroleum or when – if it comes from organisms at all.

The German chemist A. Treibs pointed out in the late forties that certain substances in oil-bearing strata are identical with porphyrins, which are decay products of chlorophyll, the green pigment of plants, and haemoglobin, the red pigment of blood. In 1965 German geochemists examined a piece of Precambrian oil shale about a billion years old. They isolated the oil from the shale and analysed it by infrared and ultraviolet spectography, gas chromatography, and other modern techniques. They found that this oil, dating back to the most primitive times of the earth, contained porphyrins in addition to a complex mixture of hydrocarbons. The composition of these compounds in the oil corresponded to their composition in living organisms of the present day. Thus the study of the shale suggests that as much as a billion years ago organisms existed which were capable of transforming carbon dioxide into food by means of photosynthesis. Another conclusion followed. Porphyrins are extremely sensitive to temperature and easily destroyed by heat. If the complicated porphyrin molecules had remained unchanged for a billion years, there could have been no periods of intense heat in earlier periods of the earth's history, as the catastrophe theorists hold. The temperatures on our planet must have remained fairly constant for at least a billion years.

Thus the Precambrian porphyrins offer evidence both for archaic life and for actualistic geology.

The Dark Chapters in the Earth's History

Otherwise, evidence for life in the oldest ages of the earth is distinctly roundabout. Since on the threshold of the Cambrian period, six to five hundred million years ago, there already existed complicated unicellular animals and plants, and in addition sponges, corals, worms, and primitive crustaceans, there must have been ancestors of these organisms in the preceding ages. A very long ancestral series must be predicated to account for the amazing evolutionary heights achieved by the most ancient fossils.

When Powell and Marsh had to leave the U.S. Geological Survey, they were succeeded by Charles D. Walcott. He divided the Precambrian into two major eras: the Archean, in which no life yet existed, and the Algonkian, in which life had arisen and gradually developed. It later became evident that Walcott's division did not suffice. The Algonkian had to be further subdivided. Life already existed, apparently, in the Archeozoic, but we know virtually nothing about it. For the Proterozoic, on the other hand, we have some fossil finds proving existence of a variegated fauna and flora.

Probably there took place in the Algonkian that momentous event which the Book of Genesis ascribes to the first day of Creation: 'Let there be light: and there was light.' In the Archean era atmospheric water vapour swathed the earth in a dense layer of cloud, cutting off the sunlight and creating a permanent dusk. But in the Algonkian the oceans had cooled sufficiently so that the sun's rays could penetrate an atmosphere saturated with water and carbon dioxide.

Little more than this can be said about the first two-thirds of the history of life on our earth. All the rest, beginning with atomic compounds and continuing on to the complicated molecular structures which led to the genesis of life, and on to the origination of the different branches of animal life, is for the present still speculation based on a variety of biological arguments but not much fossil evidence.

Such fossil evidence begins to accumulate in the succeeding time, in which palaeontologists at last begin to find solid footing: the Cambrian.

Cambrian: The Era of Trilobites

It began with the breakup of an old friendship. Two British geologists, former theology student Adam Sedgwick and former army officer Roderick Impey Murchison, had in 1830 set about investigating the oldest then known rocks and formations. Both men were friends of Lyell; both chose Wales for their starting point; both worked together so harmoniously that they seemed like a pair of scientific twins.

In the course of his labours Murchison recalled an ancient pre-Celtic tribe in Wales which had violently resisted the incursions of the Romans, the Silures. In 1835 he baptized the formation he had found after them: the Silurian. A year later Sedgwick encountered still older strata. He called them Cambrian, after the ancient name of Wales. Thereupon, inseparable as Castor and Pollux, Sedgwick and Murchison travelled throughout Great Britain and through extensive areas in France, Germany, Belgium, and the Baltic region, trying to determine the Borderline between the Cambrian and the Silurian. Each man, however, was so infatuated with the formation he had discovered that he tried to place as many strata as possible in it. Sedgwick claimed for his Cambrian, deposits and fossils that Murchison counted as belonging to the Silurian, and Murchison laid claim to as many Cambrian strata as he could for his Silurian. Finally Murchison went so far as to contend that there was no boundary; the Cambrian was really only a local aspect of the Lower Silurian. When Murchison became Director-General of the Geological Survey, most geologists accepted his dictum and restricted the use of the name Cambrian on maps.

For twenty years Sedgwick fought for his beloved Cambrian. In a voluminous work on British geology he attacked Murchison, asserting that the Silurian was not as extensive as Murchison

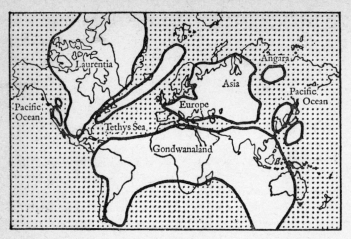

Distribution of continents and seas in the Middle Cambrian.
Based on Bölsche, Frech, Koken, Arldt, Neumayr, and Laparent.

thought. But Murchison, like Cuvier before him and Marsh after him, knew how to deal with a troublesome opponent. He finally succeeded in persuading the Geological Survey to reject all communications from his former friend and present enemy. Henceforth the word 'Cambrian' was banned from British geological circles.

In 1871 Murchison died. Sedgwick now hoped that he would encounter a better climate for his opinions, and in a scientific testament implored his colleagues to come to their senses and recognize the independence of the Cambrian system. But his hopes were in vain, and he died fifteen months after his opponent in utter despair at the blight that had fallen upon the Cambrian.

Only after his death did it become apparent that he had been quite right. The Cambrian is a real and moreover highly important period in the history of the earth. Murchison's Silurian, on the other hand, is today no longer regarded by most geologists as a unitary period. Many geologists dispense with the name; they refer to the Lower Silurian as Ordovician (after another ancient British tribe) and the Upper Silurian as simply Silurian. Thus Sedgwick won beyond the grave, like Palissy, Lamarck, and other misunderstood geniuses who could not put their ideas across during their lifetimes.

The Mysterious Three-lobed Animals

Disputes over the independence of geological formations strike us nowadays as odd and highly academic. If Lyell and the evolutionists are right, the various ages of the earth are connected intimately with one another, with barely perceptible transitions, so that it should be a matter of indifference to the geologists where the boundaries are set. In the time of Sedgwick and Murchison, however, there was a very essential difference between the Cambrian and the two Silurian epochs – at least from the palaeontological point of view. In the Cambrian no fossil vertebrates had yet been discovered; in the Silurian they already existed in the form of armoured fish, *Ostracodermi*. But this does not necessarily mean that no vertebrates existed in the Cambrian. The most primitive of living vertebrates, the lancelets, possess no hard parts that can be fossilized. Since many representatives of all other animal stocks occupied the seas of the Cambrian, it may be assumed that small, modest ancestors of the vertebrates also appeared, possibly inconspicuous organisms which showed no evident signs that their descendants would one day dominate water and land.

The really dominant creatures of Cambrian waters somewhat resembled centipedes and appeared in an incredible number of forms and sizes. Since their bodies were distinctly divided by parallel longitudinal furrows into a main section and two side lobes, they were dubbed as early as 1771 by the Diluvian Johann Ernst Immanuel Walch, *Trilobites*, 'three-lobed animals'. Walch, however, regarded them as three-lobed molluscs, akin to the present-day Venus molluscs.

Since at first only the hard back armour of the trilobites was found, not the softer ventral armour and limbs, they were ascribed to a weird miscellany of phyla. They were classed as fossil molluscs, worms, crustaceans, centipedes, spiders, and scorpions. In 1808 Pierre André Latreille, an associate of Cuvier, considered the problem. He passed a truly Solomonic judgement. If it turned out that the trilobites had had no legs, he declared, then they must be Venus molluscs; but if legs were found, then they belonged to the *Isopoda* (wood lice). The only problem was to look for legs; then the mystery of the three-lobed animals would be solved.

Charles D. Walcott, the same American geologist who later coined the term Algonkian, took up this search for trilobite legs in 1879. It turned out that many species of trilobites had the same habit as present-day isopods: for protection and in death they rolled up into balls. Walcott obtained some three and a half thousand specimens of such rolled-up trilobites and had them carefully cut open. In approximately one out of every thirteen, legs were visible. In succeeding decades William Diller Matthew and other American palaeontologists discovered a number of trilobites that were not rolled up, and on which legs and antennae could be perceived. Since the underside and limbs of the trilobites were reminiscent of the corresponding parts of crustaceans, they were generally called primitive crustaceans. But this classification is much debated nowadays. Some scientists consider them the closest relatives, or even the ancestors, of spiders and scorpions; others believe that all arthropods, from crustaceans to insects, descend directly or indirectly from trilobites.

Fascinating as Rare Stamps

As early as 1852 the Cambrian trilobites had an ardent chronicler. A French royalist named Joachim Barrande, who lived in Prague as manager of the estates of Prince Henri de Chambord, had made an extensive collection of Palaeozoic fossils from Bohemia. It included many trilobites. Barrande observed that many which looked like entirely different species were in fact only developmental stages of a single species. He was able to follow their entire growth from an almost microscopic embryo through a half dozen intermediate forms to the grown animal.

Since the days of Barrande, trilobites have exerted much the same sort of fascination upon naturalists as rare stamps upon philatelists. Their variety of forms is astonishing: there are trilobites of every conceivable size and form, mollusc-like, isopodic and crablike types, with long horns, thorns or prickles; there are some whose bodies are divided into four segments, others in

338

which up to forty-five segments of abdomen and tail can be counted. They vary even more in their eyes. Some species were totally eyeless. Others, which might also have been blind, possessed merely a single nodule-like eye. Others, again, had developed gigantic, amazingly perfected visual organs compounded of many single eyes. 'The size and number of these lenses differ exceedingly,' writes the Swiss palaeontologist Bernhard Peyer. 'Whereas the single lens may sometimes achieve half a millimetre in diameter, in other types there will be six to fourteen lenses to a millimetre.' Peyer adduces several examples of the variation in trilobite eyes: *Phacops volborthi* possesses only fourteen lenses, other types of the same genus up to two and three hundred, *Dalmanites* six hundred, *Bronteus pallifer* some four thousand, *Asaphus nobilis* about twelve thousand, and *Remopleurides radians* close to fifteen thousand.

The trilobites in the Cambrian period probably swam about the seas in incredible numbers. The blind or almost eyeless forms presumably lived deep in the mud of the bottom; the giant-eyed species may have inhabited the oceanic abysses, perhaps serving as scavengers. In the Silurian the trilobites – of unusually large size – for a time were successful competitors of other rising species, the ostracoderms, squids, and crustaceans. But then they passed the climax of their evolution and became extinct in the succeeding ages. The last of the trilobites disappeared towards the end of the Palaeozoic era, in the Permian.

The fact that trilobites were already so varied and perfected in the Cambrian served for decades as one of the chief arguments of opponents of evolution. When the Cambrian was still believed to be the oldest or second oldest period in the earth's history, the idea that any class of animals could have reached such heights by gradual evolution over so short a time was in fact difficult to conceive. But if a billion years is accorded to Precambrian life, as is now done, the trilobites had enough time at their disposal to develop into a group of animals capable of dominating the earth for a hundred million years as the saurians did in the Mesozoic and the mammals in the Tertiary period, or as man and his domestic animals do today.

Because there were so many varieties of trilobites, the animals have proved especially useful for stratigraphic purposes. In every

Palaeozoic layer there are characteristic trilobite species and groups of species which change from stratum to stratum. Thus the trilobites provide an exact picture of the succession of deposits and hence the duration of geological cycles.

Traces in Sediment

In recent times a new branch of science, palichnology or the science of fossil traces of life, has concerned itself with the trilobites in particular. Long ago Othenio Abel suspected that certain fossilized trails, called *Cruziana* traces, came from trilobites. Adolf Seilacher, the Tübingen palaeontologist, has spent almost two decades studying traces left by trilobites in marine sediments. He has ascertained that these Cambrian-Silurian animals moved straight ahead or obliquely when crawling along, but when feeding tended to move sidewise. He was able to find where they had dug, scraped, and sometimes cut deep passageways in the sea bottom.

Such traces were formerly thought to be the remains of fossil algae. Palaeontologists enumerated a large number of such 'algae', because every type of trilobite had a different digging technique, thus marking a wide variety of patterns in the petrified mud of the sea bottom: furrows, pits, tubes, filaments, stars, weaving patterns, lyre and anchor forms, even imprints that looked much like the tracks of hoofed animals and which French scientists therefore named *pas de bœuf*. Only in recent decades has it gradually been discovered that almost all these presumed 'algae' were tracks of trilobites. Seilacher even noted that the Cambrian trilobites left tracks quite different from those of Silurian and Devonian times. In rocks otherwise entirely free of fossils, trilobite tracks can therefore, as Seilacher has put it, 'also be used as time markings and thus exploited by practical geologists'.

Though much has been discovered about the trilobites, relatively little is known to this day about their contemporaries, the graptolites. These, too, first appeared in the Cambrian and became extinct in the Carboniferous. Their remains, commonly thin impressions on shale, often look like the letters of a peculiar script and were formerly considered wonders of nature, 'figure stones'. The name, graptolite, actually means 'letter stone'.

Since Cuvier's day palaeontologists have puzzled over the nature and relationships of the graptolites. Some scientists regarded them as unicellular animals related to the foraminifera, some to the cephalopods, and others to corals. Now graptolites are fairly definitely placed with a subgroup of the chordata. It is certain that the graptolites were colonial animals; they formed delicate, chitin-like shells with many cells. In each cell a single animal was housed. Dendroid graptolites were sessile (fixed) and real graptolites were epiplanktonic (attached to drifting seaweed).

The graptolite shales which incorporate most of the imprints of these mysterious creatures contain a great deal of sulphur; hence the mud layers from which this shale comes must have been so poisonous that no life could have thrived in them. It is possible that when their seaweed support sank, the graptolites died because sulphur-fixing bacteria gradually infested and poisoned the bottom.

Silurian: Fishes in Turtle Armour

Evolution, Edwin Stephen Goodrich pointed out in 1924, is frequently represented as a story of success and progress, but it is equally a story of failures and doom. Goodrich was referring particularly to a curious group of marine creatures which lived in both epochs of the Silurian and in the following Devonian. They were the oldest vertebrates whose remains have been preserved as fossils; but they looked like mistakes doomed to extinction. These strange creatures, the armoured fish, had long preoccupied the minds of scientists.

Goodrich, the last scion of a fine old English family, had shifted as a young man from art history to zoology. He became a professor at Oxford, devoting himself chiefly to the genesis of vertebrates. Under the modest title of *A Treatise on Zoology* (1909), he wrote a book destined to be a classic on fossil and living fish. It was the first great standard book on the history of fishes since the days of Agassiz. In it Goodrich summed up all the knowledge of his time; he revised the hitherto accepted family trees and sketched out a new phylogenesis which on the whole remains valid to this day.

But even this first-rate ichthyologist was unable to fit the problem of the armoured fish, the ostracoderms, into his system. What kind of organisms were they? Along with many tiny species there were some giants up to thirty feet in length. Usually all that was found of them were their heavy armour plates: petrified remains of head, thoracic and abdominal shields. Therefore they were first taken for giant water beetles, then trilobites, then gastropods, and ultimately tortoise-like reptiles.

Crab in Front and Mermaid Behind

Agassiz was the first biologist to determine that a small, especially heavily armoured type, which had been named in Lyell's honour *Cephalaspis lyelli*, had a fishlike rear part with a finned tail. As a zoologist of the time commented, it was a 'crab in front and a mermaid behind'. Consequently, in 1835 Agassiz ventured to classify these armoured creatures of the sea with the fish. 'Veritable tortoises among the fishes,' Carus Sterne later called them. In a number of species the two eye openings on the top of the head were so close together as to produce the impression of a single Cyclopean eye. In between was another opening for a 'third eye', an organ whose function no one could determine. Edward Drinker Cope studied one of these armoured cyclops, the 'winged fish' *Pterichthys cornutus*, and concluded that it had not been a fish at all, but a far older vertebrate that had managed with a single eye – a venerable link between the invertebrate *Tunicata* and the earliest ancestors of the fishes.

About 1900, however, the German zoologist Simroth turned this conclusion upside down. For in the meantime some ostracoderms with armlike forefins had been discovered; these fins even had two parts, consisting of an 'upper arm' and a 'forearm' with an 'elbow joint' between. From this Simroth concluded that the armoured fishes had been land dwellers, using their jointed forefins to crawl around on muddy banks and perhaps able to live in swamps like amphibians. From this it was only a step to the idea that the very ancient ostracoderms had been the immediate ancestors of the armoured salamanders, the first true land vertebrates. But there were also advocates of the opposite view: that the armoured fish were descended from the armoured salamanders. Was it not possible that the entire race of fish had come from four-footed amphibians by way of these armoured salamanders? The mysterious armoured amphibians and fish of the two Silurian epochs seemed to be upsetting all previous notions about the descent of the lower vertebrates.

Electric Armoured Fish

While the palaeontologists and the ichthyologists were still disputing over the nature of ostracoderms, and while highly imaginative drawings showed armoured fish basking in the midnight sun on Silurian shores, various Norwegian scientists spent some twenty years, from 1906 to 1925, collecting a large number of ostracoderm remains in Spitzbergen, Greenland, and along the Oslofjord. Most of what they found was head armour, chiefly from those relatively small cephalaspids with which Agassiz had been familiar. Erik A. Stensiö, head of the Stockholm Museum, obtained 105 specimens of cephalaspids from Spitzbergen, dissected them under the microscope with extremely delicate special instruments, and by masterly precision work was able at last to clarify the mystery of the 'turtle fish'.

The results of his studies led to a veritable scientific revolution. The cephalaspids, as Stensiö proved, were not fish at all, but cyclostomes, archaic relatives of the lampreys which exist to this day. They have no jaws like fish, but a round sucking disc for a mouth. In the structure of their brain, their sense organs, their gills and other anatomical peculiarities, the cephalaspids so closely resemble the lampreys that there can no longer be any doubt about Stensiö's analysis.

Even before Stensiö's researches most scientists had believed that the primitive ancestors of the fishes must in general features have resembled the form and mode of life of lamprey-like aquatic animals. But now it turned out that the lampreys were merely one-sided and specialized relics of an extremely ancient stock that flourished four hundred million years ago. The ostracoderms must have been the Silurian ancestors of the fishes and of the cyclostomes as well.

Stensiö's dissecting work provided a further surprise. With the aid of his excellent microscopes and his refined procedures, Stensiö was able to detect in the heads of the cephalaspids not only the network of blood vessels, but the neural network and a structure which obviously served as an electrical network. Hence the jawless ostracoderms were 'electric fish' like the present-day electric eel, capable of defending themselves against enemies by electric shocks. Presumably they used their electric charge chiefly against the six-foot gigantostracans, which looked like monstrous crabs but were in fact marine relatives of the spiders and scorpions.

Johann Kiaer, the Norwegian palaeontologist, studied another group of armoured fish, the anaspids of the Upper Silurian along the Oslofjord. These were clothed in spiny armour and likewise had a sucking disc for mouth. A third group, the pteraspids of the Upper Silurian of England and Scotland, were investigated by the English zoologist Erroll Ivor White in 1935. White found that these animals could swim very well, but nevertheless lived chiefly in the bottom ooze, sucking algae and other vegetable food out of the mud.

Did Vertebrates Arise in Fresh Water?

The jawless armoured fish seem so highly specialized that they could not possibly have arisen as late as the Ordovician. They must have had a long Cambrian and Precambrian history to evolve so far. But the older formations have yielded no fossils which fit into the ancestral series of the ostracoderms and hence of the vertebrates. Hence the distinguished American anatomist and palaeontologist Alfred Sherwood Romer conceived the thought that the oldest vertebrates may not have arisen in the sea at all, but in fresh water. Almost all the Precambrian and Cambrian sediments known so far are marine deposits. Undoubtedly inland deposits existed, but they have either not been preserved or not yet been found. Numerous fresh-water sediments have been found only in Silurian formations.

Romer examined numerous North American Silurian deposits in which remains of armoured fish had been found. Most of them, it turned out, came from fresh water. Although Romer's view has not gone undisputed, the majority of palaeontologists now incline to accept it. Vertebrates probably had their beginnings not in the oceans, but in shallow fresh-water basins, from which they gradually conquered the sea as well as the land.

The jawless armoured fish which Stensiö, Kiaer, White, and Romer studied were quite small. But they were not the only armoured protovertebrates of the epoch. Soon after them, in the Upper Silurian and especially in the Devonian, quite different armoured fish appeared, smaller and larger species, including that thirty-foot giant which was once considered a turtle-like reptile.

These placoderms, as they are called, represent the next stage in vertebrate evolution. They possessed proper upper and lower jaws. Among them were clumsy bottom dwellers which could burrow into the ooze, but also excellent swimmers which probably penetrated into the open ocean. The largest forms, the mighty arthrodires, must have been dangerous predators whose strong teeth could crack open trilobites, crabs, and smaller armoured fish. Some placoderms superficially resembled sharks, in spite of their armour. American palaeontologists therefore called them *spiny sharks*. In the Devonian the first real sharks appeared. But at the same time an entirely different group of fish developed in the fresh-water basins of the continental interiors. On muscular, stubby, finny feet these fish pushed themselves out of the water and began the conquest of dry land.

Devonian: The First Step to the Land

Only a few decades ago every zoology student learned that the swim bladder was older than the lung. The oldest fishes, he was told, had all breathed through gills. Gradually, in the course of time an air sac had developed out of the swim bladder and begun to function as a breathing organ; out of this air sac the lungs had evolved.

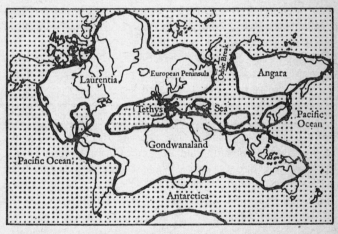

Distribution of the continents and seas in the Middle Devonian.
Based on Bölsche, Frech, Koken, Arldt, Neumayr, and Laparent.

Palaeontologists have since discovered that the process was exactly the reverse. Not the swim bladder but the lung was the original organ. All Devonian bony fishes (Osteichthytes) were dual breathers; they had lung-like air sacs along with gills. The present-day lungfishes, of which the Australian lungfish is an

347

example, are relics of that time and hence genuine 'living fossils'; they show us what the breathing organs of the Devonian fishes were like.

In fishes which remained faithful to the water, the protolung gradually developed into a swim bladder. The organ ceased to be needed for breathing and served solely as a hydrostatic mechanism. 'By alternating filling and emptying the swim bladder,' the Swiss biologist Emil Kuhn-Schnyder explains, 'fish can change their specific gravity and thus rise or drop in the water or maintain a given height without swimming movements.' Some species of fish can even produce sounds with their swim bladders. In other species of fishes, especially the deep-sea fishes, this remnant of the onetime lung has been entirely lost.

But there were also fishes in which the lung developed further and the gills gradually atrophied. That began happening more than three hundred million years ago, in the Devonian. Many detailed studies by geologists, palaeobotanists, and palaeontologists in recent decades have shown that the Devonian period must have been extremely interesting in a climatological and geographical sense – 'one of the most fascinating periods in the whole history of the earth,' Bernhard Peyer says.

Before the Devonian, dry land must have been a bare, baked, rocky landscape. But at the beginning of the Devonian a great deal of land was uplifted, and salt- or fresh-water lagoons, shallow seas, and inland waters were created. In the course of the Devonian the continents grew. Violent rainy periods alternated with periods of intense drought. The Old Red Sandstone of the Devonian formations indicates that at this time great deserts of quartz sand coloured red by iron oxides extended over vast areas of the continents. Nevertheless, the earth in the Devonian was far from 'waste and void'. Small leafless plants had clambered out on the shores of the lagoons and fresh-water lakes. All sorts of arthropods, especially crabs and scorpions, were winning a foothold on the dry land.

In the course of the Devonian the ocean at last receded so far that a great northern continent and an equally large southern continent arose, with a world-circling central sea between them. The first land plants, the psilophytes, grew into bushes and trees,

348

by the Middle and Upper Devonian forming veritable forests – although very strange forests with stems on stilts and rigid twigs that were scaly like a reptile's skin.

Lungfish and Lobefins

The Devonian takes its name from Devonshire in England. When Sedgwick and Murchison were still on good terms with each other, they recognized that the sandstone of Devonshire, with its many fossil fishes, must be a distinct geological formation. The fossilized fish were so characteristic of the Devonian that it has actually been called the Age of the Fishes. All the major groups of fishes already existed in the Devonian: sharks and other cartilaginous fish in the seas, and ganoid fish in protected basins and in fresh water – the latter the ancestors of the modern sturgeon – as well as genuine bony fish from which the majority of 'modern' fish are descended.

Of particular interest were the lungfish, which still swim in Australian, African, and South American waters. Even in the Middle Devonian strata of Scotland a large lungfish was found, *Dipterus valenciennes*, which probably fed on snails. The fossil lungfish and their surviving descendants were for a long time regarded as sterling proofs of the Darwinian theory, for they appeared to be classical transitional forms between fishes and amphibians. Their mode of breathing did not differ very much from that of some salamanders. Their larvae amazingly resembled tadpoles. Their paired fins looked almost like crude sketches of proper legs. Enthusiastically Albert Günther, the ichthyologist of the British Museum, declared that 'of all known fish fins none would be so suitable for development into the genuine leg of a higher vertebrate as that of the Australian lungfish'.

Until 1895 then, it seemed patent that the first four-footed land vertebrates had evolved out of the lungfish. But then Louis Dollo of Brussels, who had studied the iguanodons and formulated the law of the irreversibility of evolution, turned his attention to fossil and living lungfishes and to another somewhat overlooked group of fish, the lobefins or crossopterygians. Lungfish, he determined, are only a side branch of the lobefins.

Very primitive lobefins, such as *Osteolepis macrolepidotus*, already existed in the Devonian. Like the lungfishes, they possessed all the prerequisites for successful conquest of the land. Their lungs were well developed. Their fins grew from leglike stumps and could be moved by powerful muscles. Their teeth and the bones of their heads amazingly resembled those of the first amphibians.

There are a number of species of fish which would suffocate if forced to rely exclusively upon gill breathing. Their habitat is extremely polluted, oxygen-poor waters that frequently become very shallow. Hence they have developed a variety of additional respiratory methods: labyrinth breathing, intestinal breathing, storage of oxygen in special air sacs, and others.

The Devonian lobefins and their relatives, the lungfish, must have evolved in a similar way. The sharp climatic changes in the Devonian produced floods in rainy periods and extensive drying of the rivers, lakes, and ponds in the dry periods. During the drought periods the water became stagnant and poor in oxygen. Gill fish would have died by the millions in such foul waters. Fish with lungs, however, needed only to rise to the surface to obtain air. They did so not with their mouths, like other fish, but like human beings with their nostrils. They even developed the capacity for digging themselves into the mud when the water entirely dried up and waiting there for better times.

As early as the Devonian the lobefins divided into two groups, each of which took a different evolutionary path. Closely related were the lungfishes; these adjusted to life in muddy fresh water. One group of lobefins went into the sea; they moved first into sheltered lagoons and then into the deeper regions of the ocean. In the Mesozoic the chief representatives of this group, the coelacanths, flourished for a long time. The classic fossil sites in Sussex, Solnhofen, Holzmaden etc. yielded numerous coelacanth fossils.

In 1938, however, scientists and through them the world public were astonished to learn that one species of these fossil fish, *Latimeria chalumnae*, is still among the living. Since then a number of specimens of surviving coelacanths have been caught in the sea to the north of the Malagasy Republic. The first press reports were somewhat reckless in calling this 'fish with legs' a direct ancestor of all higher vertebrates including man. This is hardly accurate. Like the fossil coelacanths, the living *Latimeria* belongs to a side branch of the lobefins. These marine crossopterygians developed along a parallel line to the line of the modern bony fishes; they abandoned lung and nose breathing in favour of gill breathing, in order to meet the competition of other fishes which were far better adapted to life in the water.

Of far greater importance to the further history of life is the second group of lobefins, which descended directly from the Middle Devonian *Osteolepis*. Many years of research by William King Gregory, Edwin Stephen Goodrich, D. M. S. Watson, Alfred Sherwood Romer, and T. Stanley Westoll have demonstrated that the four-footed vertebrates indeed descended from this third group. We men are ultimately as much descendants of the osteolepid lobefins as the salamanders.

In his book *Man and the Vertebrates* Romer described the great event of the Devonian period, the first step up to land:

'The most primitive of known amphibians were ... inhabitants of fresh-water pools and streams in Carboniferous and Devonian times. Alongside them lived representatives of the ancestral crossopterygians, forms similar to them in food habits and in many structural features and differing mainly in the lesser development of the paired limbs. Why should the amphibians have developed these limbs and become potential land dwellers? Not to breathe air, for that could be done by merely coming to the surface of the pool. Not because they were driven out in search of food, for they were fish-eating types for which there was little food to be had on land. Not to escape enemies, for they were among the larger animals of the streams and pools of that day.

'The development of limbs and the consequent ability to live on land seem, paradoxically, to have been adaptations for remaining in the water.'

As long as rivers and ponds were full of water, the lobefins, being the better swimmers, had better chances for survival than the amphibians. But the devastating droughts of the Devonian repeatedly dried out almost entirely the shallow fresh-water lakes. The remaining water became stagnant. In these conditions lobefins and amphibians had the same chances, for they could both rise to the surface and breathe air. But sometimes the Devonian droughts grew so intense and lasted so long that the water entirely disappeared from lakes and ponds. Then the four-footed amphibians had the advantage. The lobefins could only crawl into the mud and wait for rain; they might also make their way to other bodies of water on their incipient legs, if the water were not too far away. But the amphibians were capable of moving a considerable distance overland in their search for water. 'Once this development of limbs had taken place, however, it is not hard to imagine how true land life eventually resulted,' Romer comments.

Since the Carboniferous swamps and rain forests swarmed with primitive dragonflies, centipedes, scorpions, spiders, daddy long-legs, and other arthropods, the amphibians found enough food on land. It is therefore not surprising that the armoured batrachians soon grew into giant forms the size of crocodiles. By the end of the Carboniferous the first primitive saurians emerged from them.

Carboniferous: From Club Moss to Coal

Even the ancient Chinese, Greeks, Romans, and other civilized peoples of the past knew the value of coal. But until recent times no one was aware of what the black, hard, combustible substance really was.

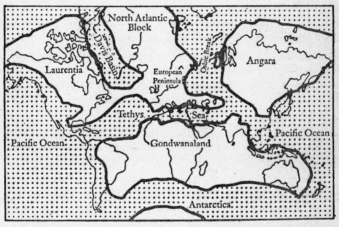

Distribution of continents and seas in the Upper Carboniferous. *Based on Bölsche, Frech, Koken, Arldt, Neumayr, and Laparent.*

The great German mineralogist Georg Bauer, who called himself Agricola, decided in 1544 that coal must be petroleum which had mixed with sulphur and become hardened by long storage in the ground. A century and a half later Johannes Bünting, a mining engineer, reported that there were two theories about the origin of coal. According to the first, God had created coal at the beginning of the world; according to the second, it arose later 'by the power of nature and the force of the earth'. The first view was correct, he declared, the second heretical and hence despicable. There could be no doubt that coal had existed since the beginning of the world. Therefore there was no danger that the human race would ever run out of coal because coal was provided with 'special seeds for its reproduction and growth under the ground'.

Agricola's view that coal was condensed petroleum was vigorously championed by a number of naturalists as late as the nineteenth century. The English Diluvian William Buckland toyed with the idea. The Irish Neptunist Kirwan argued in 1812 that coal seams were merely decay products of primitive rock. As late as 1844 Andreas Wagner (the man who described the apes of Pikermi and the first bird) held that the superfluous carbon dioxide in the air precipitated on earth, lost its oxygen and condensed into coal.

The first man to speak up for a vegetable origin of coal was that old Diluvian Johann Jakob Scheuchzer. In 1706 he contended that peat, lignite, and coal were only various stages on the way to the carbonization of vegetable matter. No one, Scheuchzer emphasized, doubted that peat was of vegetable origin. The connexion between peat and lignite was equally obvious. Consequently, one need only go one step further and consider coal as petrified vegetable remains – the compressed remains of ancient peat bogs together with other plants that had been washed into them by floods.

Scheuchzer's idea was taken up by other scientists only towards the end of the eighteenth century – with some amendments. In Germany Count von Beroldingen, in France the great botanist Bernard de Jussien, and in Scotland James Hutton, founder of the Plutonic-Vulcanian school, declared emphatically that coal must have arisen from plants – not from ancient bogs, as Scheuchzer had thought, but from primordial forests.

The Greatest Forests of All Time

In 1848 the German botanist Heinrich Robert Göppert studied a number of thin cross sections of coal under the microscope. He was able to demonstrate the presence of plant cells in the coal. Göppert then applied heat and pressure under water and produced both lignite and coal artificially, out of a variety of plants. Thus it became unequivocally clear that coal was the fossil remains of primitive flora.

But what plants had formed coal in the Palaeozoic era? This, too, remained a fiercely debated question for many years. In 1808 the Belgian geologist Omalius d'Halloy conferred the name Carboniferous, 'coal-bearing', upon the geological system in which coal seams are particularly common. It was not yet known that in the preceding Devonian strata and the following Permian strata, as well as in Mesozoic strata, coal is found. A number of palaeontologists and geologists returned to Scheuchzer's ideas, insisting that the coal must have been formed in enormous peat bogs; others, also following Scheuchzer, held that coal must have formed from tree trunks washed by the floods into shallow sea basins.

One theory urged with particular stubbornness held that coal had derived not from trees or other land plants, but from seaweed and similar marine vegetation. Professor Göppert had to deal with all these arguments and hypotheses. He undertook chemical analyses to show that coal could not have been formed from marine plants, but only from land plants. Further microscopic studies proved that trees of varying sizes must have been involved, chiefly great ferns and coniferous trees. The horizontal layers of coal seams contained, he discovered, many fossil tree trunks standing upright. He thus came to the conclusion that at one time in the Carboniferous there had existed 'luxuriant forests of cryptogams whose dead remains fell to the ground and in shelter of the water underwent the carbonization process'. This view was violently opposed as late as the end of the last century, but it is now generally accepted.

The Carboniferous coal stocks are the remains of the largest stock of flora ever to have existed on our earth. In the Carboniferous period animals did not yet dominate on land. The land was the kingdom of plants, of swamps thick with undergrowth, of dripping wet, mist-drenched, luxuriant rain forests. Everywhere on earth, even in the Arctic and Antarctic regions, the ground had become a fertile, steaming morass. Plants grew lushly on every square inch. They died, dropped into the mud; the tree trunks crashed to the ground; their decay products fertilized the soil and provided humus for countless succeeding generations of plants.

Most of the red quartz sand deserts of the Devonian vanished under a green carpet and were transformed into moist, fertile soil. Only a few desert and high mountain regions remained bare and lifeless. The Carboniferous climate was primarily responsible for this striking change in the whole aspect of the earth and the tremendous upsurge in plant growth and evolution. No one knows why the Devonian was so dry and the Carboniferous so wet. Cosmic events have frequently been invoked as explanation. But probably the development of the plants and the climatic changes went hand in hand. When the plants had conquered the land, they attracted the rain, so to speak. Regions with heavy vegetation, tropical jungles for example, are to this day extremely rainy. It is not necessarily the case that the rains produce the jungles; the reverse may be true. The damp laundry-room atmosphere which thus arose in the Carboniferous further promoted plant growth, which in turn stimulated more rainfall. Thus in the course of fifty million years virtually the whole land area of the earth became one vast tropical jungle.

The cryptogams of which Göppert spoke are spore plants; present-day ferns, horsetails, and club mosses are cryptogams. In the Carboniferous, as Göppert pointed out, they grew into gigantic trees. Close relatives of the club mosses, the scale trees became the characteristic plants of the Carboniferous in many regions and provided the chief material out of which coal was formed. In other Carboniferous forests, gigantic horsetails predominated; like mangroves they stood on stilt-like roots in the mud.

Aside from these there were also very curious-looking trees which seem exceptionally 'modern' compared to the primitive cryptogams. These stood on slender trunks rising some hundred feet in the air to a palmlike crown. Their ribbon-like leaves could attain a length of three feet and more. When these *Cordaites* bloomed, the flowers looked like ears of grain. The *Cordaites* were not, in spite of their appearance, ancestors of the palms but of a type of plant destined to be far more successful – the coniferous trees which were to attain dominance in following eras.

The fauna of the Carboniferous seems to have taken second

place. However, this period saw the rise of the phylum with the most numerous species in the animal kingdom, the arthropods. Along with spiders and millipeds, innumerable hordes of insects swarmed about the Carboniferous vegetation. Dragonflies grew to immense size, with wingspans of up to two and a half feet. Perhaps primitive forms of most groups of present-day insects existed even then, for the swampy forests offered them ideal conditions of life. But insects are preserved in the fossil state only under exceptional circumstances. Nevertheless, about a thousand species of insects from the Carboniferous have been identified.

The swamps were inhabited not only by large labyrinthodonts, but also by other amphibians, including the ancestors of frogs and salamanders. At the transition between the Carboniferous and the Permian a group of amphibian-like creatures appeared who were the joy of evolutionists because they seemed to constitute ideal connecting links. These are the cotylosaurs.

'They still possess an amphibian-like skull,' writes Kuhn-Schnyder. 'In their manner of movement they can scarcely be distinguished from the old amphibians. In skeletal structure, too, the boundaries between the early amphibians and the cotylosaurs is blurred. No wonder, then, that the position in the system of certain finds is uncertain.'

One such 'uncertain' animal is *Seymouria baylorensis* from the Red Beds of Texas. The discoverer of this fossil, the American palaeontologist T. E. White, called it an 'archaic form, the most primitive reptile we know today'. Kuhn-Schnyder, however, comments: 'The skull shows close kinship with fossil amphibians. ... On the basis of skeletal structure it is impossible to decide whether *Seymouria* should be classed with the amphibians or the reptiles. Did it lay its eggs in water or on land?'

Such transitional forms are particularly fascinating to palaeontologists. Amphibian or reptile – that was the question on the boundary between the Carboniferous and the Permian. One period later, on the boundary between the Permian and the Triassic periods, the question was: reptile or mammal? For the age of modern life dawned in the Permian.

Permian: Boundary Stones Between Two Worlds

Travelling in Russia back in the 1840s, Murchison had observed that the system of rocks following the Carboniferous was particularly well developed in the Russian province of Perm. He therefore named this time-span the Permian. There is no clear dividing line between the Carboniferous and the Permian, and to this day scientists dispute over where the one ceases and the other begins. Nevertheless, in all of geological history there can be scarcely greater contrasts than those between the Carboniferous rain forests with their hothouse atmosphere and the Permian clayey deserts with their erosion, their volcanic eruptions, and their periodic ice ages.

The Permian ice ages affected only the great southern continent, Gondwanaland. There the swamp forests vanished, to be replaced by sandy deserts, inland seas, and extensive deposits of salt. In both hemispheres many of the characteristic creatures of previous ages died out: the trilobites, graptolites, scale trees. Others continued on, but played a more modest part: the lobe-fins, giant batrachians, ferns, and horsetails. Others, again, flourished, and gave hints of their future dominance: ammonites, reptiles, palm ferns, and coniferous trees.

Perhaps there is a link between the rise of coniferous forests and reptiles and the formation of deserts in Permian times. Plants and animals now had to be hardy enough to survive even in dry regions.

South Africa was particularly affected by the Permian Ice Ages. In the region of the great Karroo Desert there are many rocks which are startlingly reminiscent of the erratic boulders of our Pleistocene Ice Age. Above these rocks, which bear the characteristic parallel scratches of glacial movements, lie thick strata of sediments, some of which belong to the Permian, some to the

Triassic. Indications are that the Karroo Desert was at the time not particularly rich in life. Ice, drought, and flood had decimated plants and four-footed animals. One group of reptiles, however, appears to have thrived in this climatically unpleasant period. These were the theromorphs, the mammal-like saurians.

The oldest representative of this saurian group, precursor of the mammals, was found in the Joggins formation of Nova Scotia. Families and genera of more recent age were then discovered elsewhere in North America, in South America, Eurasia, and Africa. Among them were some very curious forms: giants and dwarfs, crude creatures with turtle-like horny beaks, carnivorous and herbivorous reptiles with sail-like structures or a multitude of spinous processes on their backs, predatory living dragons with terrifying rows of teeth, clumsy toadlike cave dwellers, and even amphibian-like saurians which had the mysterious organ known as a pineal eye on their heads.

In the Permian the saurians for the first time developed in an amazing variety of sizes and shapes. The Permian reptiles were an advance guard, as it were, for the mighty hosts of giant saurians which were to take over in the following period, the Triassic. But these many and varied saurians with the exception of the South African types did not lead to the mammals. The American and Eurasian theromorphs were only *like* mammals; they were, as Edwin H. Colbert has put it, 'aberrant mammal-like reptiles', special types characteristic of the Permian transitional period, and soon to vanish again. Only in the Karroo region were things otherwise. Here a few Permian reptiles flourished which in structure of jaw and teeth astonishingly recall primitive carnivorous mammals. The large canine teeth looked like the fangs of predatory mammals. The spines, shoulder blades, hipbones, limbs, and feet also showed many characteristics typical of the mammalian structure.

Mammalian Ancestors

Robert Broom, the son of a textile pattern designer, discovered his passion for natural history when he played on the seashore as a small boy in Scotland. He studied anatomy, became a physician, and emigrated to Australia to start a practice and on the side investigate the primitive marsupials of the island continent. Although professional duties took up most of his time, he succeeded in finding a large number of fossil marsupials from Tertiary and Quaternary strata of Australia. With these discoveries behind him, he grew more and more eager to search for the real ancestors of mammals.

In 1896 he heard that in Cape Colony, South Africa, some mammal-like saurian fossils had turned up, theriodonts, as they were called. Broom promptly set out for South Africa and settled down as a country doctor on the fringes of the Karroo Desert. He utilized his every leisure hour to go out into the desert and dig for fossils. In the course of his endeavours he dug up so many mammal-like reptiles that Henry Fairfield Osborn, whose attention had been drawn to this remarkable amateur, invited him to New York in 1906. Osborn encouraged Broom to continue his excavations, and bought many of his finds for the New York museum. Ten years later, Broom was at last able to clarify the origin of mammals.

Huxley had once suggested that mammals arose directly from amphibians. According to Cope, the true ancestors of the mammals were the theromorphs. Broom confirmed Cope's view, but discovered that a special group of South African theromorphs, the ictidosaurs, were already half-way to mammalian status. In the South African Permian and Early Trias, he demonstrated, there had been a considerable number of mammal-like saurians, among them *Cynognathus*, *Lycaenops*, *Bauria*, and others. In the Upper Trias these forms passed into ictidosaurs almost without a break. Animals such as *Thrinaxodon* and *Sesamodon* were 'mammal-like reptiles par excellence . . . the ancestors of our own blood kin'.

A reptile skull, as every zoologist knows, differs from a mammal skull in other characteristics besides dentition. There are fossils with nearly mammalian teeth which nevertheless must still be classed among the reptiles, and fossils with reptilian teeth which already belong among the mammals. The real criterion for distinguishing a reptilian from a mammalian skull is the structure of the lower jaw. If the lower jaw consists of several elements, the animal must be a reptile; if it is formed of one piece, it belongs to a mammal. In Germany, England, China, and Africa fossil teeth have been found which look so thoroughly mammalian that no one doubted they belonged to some primitive mammal. But when the jaw belonging to these teeth was found, it proved to be divided into three parts. Was the animal therefore a reptile?

Broom's finds in South Africa shook the seemingly solid foundations of this distinction. For the ictidosaurs formed an intermediate stage between the reptilian and the mammalian lower jaw. In the oldest forms the jaw consisted of several independent bones; in the most recent forms all but the dentary had become incorporated into the ear mechanism or had been lost. Clearly, the ictidosaurs stood on the very dividing line between reptiles and mammals.

'Finding the Missing Link'

Most Permian-Triassic saurians stopped half-way in their development towards mammals. Only a few very small forms succeeded in maintaining themselves during the Mesozoic Age of Reptiles. These triconodonts, pantotheres, trituberculates, and multituberculates of the Triassic, Jurassic, and Cretaceous have left behind only fragmentary teeth and remains of jawbones, so that it would be vain to undertake complete comparisons with the teeth and skulls of present-day mammalian groups. They are simply referred to as 'Mesozoic mammals', and few attempts have been made to guess what they may have looked like and where they fit into the system. Presumably the duck-bills, the marsupials, and other placental animals sprang from different roots in the Mesozoic era.

Broom belongs among the few palaeontologists of our century who received any official recognition from his country. In general, politicians manifest a staunch indifference to palaeontological discoveries, and often prove hard of hearing when scientists need funds for such purposes. But Broom found a faithful friend and supporter in Jan Smuts, the last democratic Prime Minister of South Africa. Smuts freed him from the necessity of earning a living as a country doctor by appointing him curator for palaeontology at the Pretoria Museum.

The discoverer of the transitional forms between the reptiles and the mammals closed another gap that is of utmost interest for us human beings. Up to his death in 1951 Broom remained a specialist on connexions between the different stocks of animals. One of his books bears the arresting title *Finding the Missing Link*. By 'missing link' Broom did not in this case mean a mammal-like reptile, but a group of half-ape, half-human creatures whose remains he had blasted out of the limestone caves of South Africa. These *prehominids* were perhaps the most important fossil finds of recent decades. They were important for anthropologists, philosophers, and theologians as well as zoologists, for they established the long-sought bridge between animal and man whose existence Darwin had predicted.

Triassic: Tracks in the Sandstone

The analysis and interpretation of primordial tracks is a science in itself, perhaps the branch of palaeontology closest to detective work. Innumerable creatures of the ancient world have left their tracks in mud, sand, or clay. By piercing together many bits of evidence it becomes possible to visualize what manner of creature must have made the tracks. With such a set of portraits, scientists can fill in whole chapters on the life of past ages.

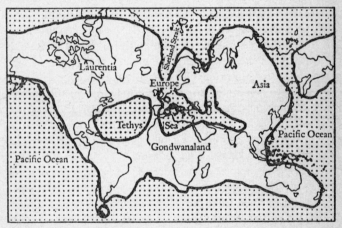

Distribution of the continents and seas in the Middle Triassic. *Based on Bölsche, Frech, Koken, Arldt, Neumayr, and Laparent.*

The ocean ooze has preserved the marks of trilobites, sponges, worms, mussels, sea urchins, and crabs. On the bottoms of rivers, brooks, and fresh-water lakes the trails and burrows of antediluvian snails, fishes, and tadpoles can be seen. On damp beaches and on dry land are the footprints of amphibians and saurians, birds, carnivores, giant sloths, ungulates, and mammoths.

Sometimes such vestiges sow confusion, as did a number of fossilized tracks from Nevada which confounded the anthropologists as well as the palaeontologists. In 1882 a new city jail was to be built in Carson City. The sandstone for the building was taken from a nearby quarry of Pleistocene rock and clay strata which proved to be a treasure trove of primeval footprints. The sandstone slabs bore unmistakable tracks of elephants, stags, horses, wolves, and birds; but there were others that looked amazingly humanoid.

The palaeontologists who studied these mysterious footprints agreed that they were made by an animal with a large wide sole whose toes either did not press into the ground at all, or if they did so, then in a highly unusual manner. But what kind of animal could that be? At first there was talk of a large bear. Then such palaeontological leaders as Marsh flirted with the idea that all existing theories and all previous researches to the contrary, there might after all have been Pleistocene men in America. A man named Loudenback developed a theory of human origins which culminated in the argument that human history may not have begun in the Old World, but in America.

Finally the palaeontologist Chester Stock hit on the correct interpretation in 1925. Subscribing for a while to the bear and to the human theory, he then made systematic studies and showed that a sloth, probably a giant mylodon or nothrotherium, must have left the Carson City imprints.

The best known, most discussed, and most mysterious of fossil trails, with which a great many prominent palaeontologists dealt for a century, come from the Triassic rocks. They look almost like the petrified imprints of a human hand. The unknown author of the prints was therefore dubbed *Chirotherium*. The unmasking of the chirotherium belongs among the great coups of palaeontological detective work.

Handprints on Red Stone Slabs

We have lost trace of who may first have seen these 'petrified hands'. Tracks of chirotheres have been found in so many places in Germany, England, France, Spain, and North America that examples must surely have come the way of the collectors of fossils by the Renaissance. But their importance went entirely unrecognized; for it was only in 1824 that an English amateur geologist had the idea of saving one of the slabs, which he had found at Tarporley in Cheshire, and showing it to a specialist named Grey Egerton.

At first Egerton could make nothing of the tracks. Fifteen years passed before he finally decided to publish a report 'on the imprints of the hind feet of a gigantic animal from the New Red Sandstone of Cheshire' in the *London and Edinburgh Philosophical Magazine*. Within the same issue Mr Cunningham described similar tracks that he had found at Storeton, likewise in Cheshire. The reason these Englishmen were suddenly publicizing their discoveries lay outside the borders of Great Britain. They were responding to two developments in Germany: a new period had been added to the recognized periods of geological history, the Triassic; and the Germans had come upon splendidly preserved chirothere tracks which were arousing excitement everywhere in the scientific world.

The formations between the Permian and the Jurassic had hitherto been neglected by English geologists. Smith, Sedgwick, and Murchison had used the name New Red Sandstone for the age which we today call the Triassic and place some two hundred million years ago. They had not paid much attention to it because there were few obvious Triassic formations in England and few Triassic fossils. On the Continent, on the other hand, this first period of the Mesozoic era has left conspicuous traces. Coloured sandstone, shell limestone, and red marl, the three classic formations of the Triassic, gave the Black Forest and the Vosges their characteristic appearance. And since Friedrich von Alberti, a mining engineer and salt mine inspector, assumed that the early Mesozoic layers must be divided into these three strikingly distinct stages everywhere in the world, he designated a special period to be named Trias, 'three-stage time'.

Later, it turned out that this tripartite division applied only to Central Europe. The shell limestone, a typical marine deposit, is found almost exclusively in Germany, for northern Germany was covered by an inland sea in the Middle Triassic. In most other

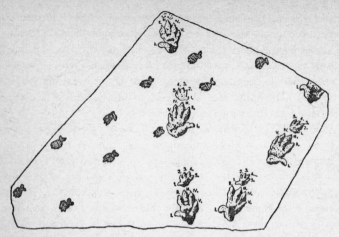

Tracks of a smaller species of *Chirotherium* from Scotland.

regions of the globe the Triassic consists only of two formations which are comparable to the coloured sandstone and the red marl of Germany. Nevertheless Alberti's name was retained, geologists having grown so accustomed to it.

Aside from the German shell limestone formation, this period was a continuation of the dry and frequently barren Permian. In the Triassic the continents attained the greatest extension in the history of the earth. Vast new land masses appeared in the northern and the southern hemispheres. Much of this land was arid desert. Since the sand was often dyed a deep red by iron oxides, the sandstone rock that eventually formed out of it has become a prized building material for cathedrals, palaces, castles, and fine houses. The cathedral of Strassburg, the palace at Heidelberg, many of the fine buildings of Worms and Speyer, and, indeed, some of the most famous Gothic and Renaissance structures were made of this red sandstone from the Triassic deserts.

In the Triassic the empire of reptiles on water and land had already reached its height. Virtually all mammalian groups were likewise present. One of these played a significant part in the history of the chirothere. They are the *Pseudosuchia*, 'false crocodiles', as they are misleadingly called; for they had nothing to do with crocodiles. Some were biped pseudosuchians which were close to the ancestors of the birds. Other species walked on all fours or at least used their hands for running and jumping. Since most pseudosuchians were probably desert animals, scant remains of them have been found. For in deserts eaters of carrion make short work of the corpses of animals, including the bones. On the other hand, when a rare pseudosuchian was found in the sandstone or the red marl, it was very likely to be in a fine state of preservation. Sand and the heat of the sun would have converted it into a lifelike mummy.

The same year that Alberti coined the term 'Triassic' a German geologist, F. K. L. Sickler, found a number of footprints in the dark red sandstone of Hildburghausen which were exactly like the fossil tracks of Cheshire. Sickler was not familiar with the English finds. But he, too, was astonished at the resemblance of the Hildburghausen tracks to the imprints of human hands. He therefore appealed to the aged Professor Blumenbach in Göttingen, describing to him 'the extremely curious reliefs of tracks of large, unknown primordial animals discovered only a few months ago in the Hessburg sandstone quarries near the city of Hildburghausen'. He also published an account of his findings.

In the wake of this, English scientists were prompted to come forth with their finds. Nor were they the only ones, for before the year was out, no less than nine other papers appeared in Germany on the mysterious 'hand animal'. The opinions of these authorities diverged enormously. Because the apparent hand prints looked so astonishingly humanoid, Friedrich S. Voigt held that the animal must have been a great ape. He even gave the ape a name: *Palaeopithecus*. Later Voigt realized that this hypothesis was too audacious. He concluded that the large tracks came 'from the famous cave bear', the smaller ones 'probably from a mandrill'.

Voigt overlooked the fact that apes or bears could not yet have been present in the Triassic. Two other authorities who really should have known better likewise reasoned wrongly: the palaeontologist Johann Jakob Kaup, and the versatile genius Alexander von Humboldt. Both concluded that the maker of the tracks had been a large marsupial. Kaup gave the unknown creature the name it still bears, *Chirotherium*, which really means 'hand mammal' – for zoologists generally apply *therium* only to mammals. Alexander von Humboldt at first supported Kaup's view; later, he decided that it could not be settled so rashly. Although he too regarded the chirothere as a mammal, he proposed that it be named provisionally either *Chirotherium* or *Chirosaurus*, thus holding open the possibility that the prints had come from some saurian.

The Mysterious Thumb

In the following decades more chirothere tracks, large and small, were reported from the most varied regions of the earth. In 1857 the French geologist Gabriel Auguste Daubrée published a report on 'traces of quadrupeds in the sandstone of Saint-Valbert near Luxeuil (Haute Saône)'. Daubrée also delivered the judgement that the tracks had been made by a mammal. He even analysed the skin structure which he thought he discerned in the prints, and emphatically supported the opinions of Kaup and von Humboldt.

More and more chirothere finds were made in England, Germany and France. But it was always these typical familiar prints found in the sandstone; no one came upon a single remnant of the mysterious animal. In 1898 Calderón found the first Spanish chirotheres. By 1900 some eight kinds of chirotheres had been discovered in North America, particularly in the states of Colorado, Arizona, and Utah, with their vast stretches of desert.

By this time no one believed that apes, bears, marsupials, or other mammals could have made the prints. Scientists had split into two parties, the debate being whether the chirotheres had been amphibians or reptiles. For a while the idea of amphibians seemed to be winning favour. As early as 1835 the German geologist H. F. Link had described the Hildburghausen prints as the hopping marks of a gigantic toad. Six years later Sir Richard Owen declared that they could only have been made by labyrinthodonts, the large armoured amphibians. Most scientists in the nineteenth century agreed with Owen. Lyell, too, who so often took a view opposite from that of Owen on many other questions, concurred completely in the case of the chirothere. He even tried to reconstruct the animal from its tracks. In the drawing he presents a toadlike labyrinthodont running cross-wise by setting its right hindfoot down on the left and its left on the right.

Lyell's 'chirothere', a labyrinthodont running crosswise, thus making the 'hand prints'.

Bizarre though Lyell's drawing appeared, there was good reasoning behind it. The chirothere tracks were five-toed; but the outspread big toe, which looked so very much like a thumb, was on the outside rather than the inside. If this large fleshy toe or thumb actually were that, the chirothere must have walked crosswise. This kind of motion, Owen's and Lyell's followers believed, was far more likely in an amphibian than in a reptile.

Some, however, disagreed sharply. Why could a reptile not have walked in the same odd way? they asked. Among these critics was the British scientist Williamson, who in 1867 proclaimed the chirothere a primitive crocodile. Twenty-five years later Albert Gaudry called it a dinosaur. Thereafter the dinosaur idea persisted alongside that of labyrinthodont. But neither could be proved.

In 1917, by which time chirothere research seemed to have come to a dead end, Karl Willruth, a young German palaeontologist at the University of Halle, devoted his doctoral thesis to the mysterious prints. Fortunately Willruth had not read the innumerable ingenious articles that had been published since the time of Lyell and Owen on crosswise gaits and outside thumbs; or if he had read them, he disregarded them. But on the other hand he had come across a forgotten old article by Croizet dating from 1836 and a still unpublished essay by the geologist Bornemann of 1889. In both these papers Willruth found confirmation of his view that the chirothere had not walked crosswise, but quite normally. The mystery of the 'outside thumb' was explicable in a very simple way according to Willruth; the supposed 'thumb' had not been a toe at all, but merely a fleshy appendage.

When Willruth's professor, Johannes Walther, read the dissertation, he did not reprimand the young man for his heretical views, but congratulated him for having solved the problem. By this time, close to sixty books and articles had been published on the chirothere. Almost all the well-known geologists and palaeontologists had offered opinions on the question of the hand animal, with the result that it was the subject of more general confusion than almost any other find. Professor Walther was delighted that someone had found a way out of the dilemma.

Willruth's idea was taken up by a Hungarian who can be regarded as the real founder of the discipline of fossil tracks. Baron von Nopcsa was interesting in another respect, for he was one of the most restive personalities in the history of palaeontology.

Baron Franz von Nopcsa, a nephew of the Chief Steward of the Empress Elisabeth of Austria-Hungary, came from old Hungarian-Transylvanian nobility. His interest in palaeontology dated from his eighteenth year. His sister Ilona had found a number of strange large bones in the park of Szacsal, the family estate. Since the young man happened to have business in Vienna, he took the finds with him and showed them to the leading geologist of the capital, Eduard Suess. Professor Suess explained that these were dinosaur bones from Cretaceous times, and asked him to appear before the Vienna Academy to speak on the discovery.

The young dilettante quailed. He did not know enough, he replied. But Professor Suess, who must urgently have needed scientific apprentices, replied harshly: 'If you don't know enough kindly learn!' and ordered the young nobleman to attend his lectures. Thus the baron became a geology student, and later the leading geologist and palaeontologist of Hungary.

Before long Baron Nopcsa enjoyed great prestige throughout the world in herpetology, the science of reptiles. But it went against his grain to devote himself entirely to research, for he was a born adventurer. In fact, shortly after he completed his university studies he travelled in the Balkans as a secret agent for his government conducting his intelligence work behind the screen of research. He also lived in Albania disguised as a Skipetar, an Albanian mountaineer. He was adopted into the tribe of the Mirdites, mastered the extremely difficult language until he spoke it like a native, and in 1913, when Albania was liberated from Turkey, he considered that his great hour had come. He hoped to be made King of Albania. But the great powers preferred to place on the throne the irresolute Prince Wilhelm zu Wied, who had no knowledge of the country and had to leave it after only six months. After the end of the First World War, Achmed Zogu, later dictator and King of Albania, crushed a revolt of the Mirdites, destroying for good Baron Nopcsa's dreams and machinations.

It is hard to think of so spirited and ambitious a personality following the sober path of a palaeontologist. To Baron Nopcsa, however, a scientific discipline seems to have provided the necessary balance in his otherwise chequered and perilous life. With his active mind fertile in new ideas, he reanimated palaeontology in many fields. In 1923, just at the time he had to renounce his soaring Albanian projects forever, he published a work on the tracks of different types of reptiles and amphibians. From the position of the leg, the depth of the print, the angle formed by the limbs with the ground, and several other components, he was able to distinguish different groups of tracks and ascribe them to specific groups of animals: labyrinthodonts, primitive crocodiles, dinosaurs, lizards, and others. He assigned the tracks of the chirothere to the dinosaur group. The supposed thumb was, as Willruth had argued in his dissertation, 'only a partly fleshy appendage'. And the chirotheres themselves, Nopcsa reasoned, must have been fairly close to a genus of large Triassic dinosaurs, the plateosaurs.

Since Baron Nopcsa was by this time the world's greatest expert in palaeontological tracks, other scientists might have been expected to accept his observations and start looking for suitable plateosaurs whose footprints agreed with the 'chirothere' tracks. But in the one university town where the largest number of plateosaur finds had been made, and where the foremost specialists in plateosaurs could be found – Tübingen – his colleagues had serious doubts about Baron Nopcsa's view.

Trossingen – Caravan Route for Dinosaurs

The first plateosaurs were discovered in 1909 by workmen in a brick-clay pit near Halberstadt. Two years later Eberhard Fraas, the son and successor of Oskar Fraas, found a still greater source of them near the Swabian village of Trossingen, in the Rottweil area. From 1921 to 1923 Professor Friedrich Freiherr von Huene of Tübingen undertook extensive excavations there, and un-covered bones in quantities comparable only to the great 'fossil graveyard' finds of North America.

Professor von Huene learned why the plateosaurs of Tros-singen had been so splendidly and completely preserved. When the animals died, they were immediately covered with the red desert sands. In the Upper Triassic period there had been a virtual 'caravan route for dinosaurs' in the vicinity of Trossin-gen. The animals had apparently taken this same route at re-current intervals. Professor von Huene commented: 'I assume that at the beginning of the dry season they moved in hordes from the mountains down to the coast, to return to the mountains before the beginning of the rainy season.'

Since the dark-red rocks of Trossingen date from the Upper Triassic, the plateosaurs must be far younger than the chiro-there tracks. But the conditions of life in the red marl time of the plateosaurs resembled those in the red sandstone time of the chirotheres. Baron Nopcsa's thesis that ancestors of such desert migrants as the plateosaurs might have left the mysterious chirothere marks therefore seemed plausible. For in the earliest epoch of the Triassic dinosaurs also made regular migrations across the desert. Of course the makers of the chirothere prints must have been considerably smaller saurians than the very large plateosaurs.

Soon enough, in fact, a number of characteristic tracks were found that could have come from plateosaurs or from the closely related bipedal dinosaurs. These turned up not in Trossingen, but in the red sandstone of Connecticut and Massachusetts. Saurian experts in Tübingen, however, promptly noted that these prints were utterly different from those of the chirothere 'hands'. Plateosaurs were five-toed, but only three of their toes touched the ground as they walked. These three clawlike toes dug in deeply, producing tracks almost like those of a gigantic bird.

The Tübingen scientists therefore refused to believe that the chirotheres could possibly have been identical with plateosaur-like animals of the Lower Triassic. Indeed, one palaeontologist at Tübingen University, Wolfgang Soergel, had long ago begun on another tack entirely.

The Chiniquà Saurian

Soergel carefully studied all available chirothere tracks, especially those of the large *Chirotherium barthi* which occurred with particular frequency in South Germany. He concluded that Willruth and Baron Nopcsa had been mistaken about a crucial point: the apparent 'thumb' was *not* a fleshy appendage after all. It showed distinct jointing and therefore must be a genuine toe, but not the first. Rather, he reasoned, it was the fifth or 'little' toe, which in the chirotheres was not little at all, but unusually large and, moreover, set at a wide angle from the other toes.

Which of the reptiles had possessed a large fifth toe set at a wide angle from the others? None of the plateosaurs and their relatives. Soergel patiently went through all the collections once more. Finally he found a promising group. They were certain *Pseudosuchia* of the South African Triassic, *Euparkeria capensis* and other species. But alas, the size of the pseudosuchians did not correspond with the size of the chirothere tracks. Some chirothere 'hands' were more than eight inches long; the foot of *Euparkeria* was barely two inches long. Whereas the plateosaurs were much too big, *Euparkeria* was much too small.

Nevertheless, Soergel published a study entitled 'The Tracks of the Chirotheria' (1925) in which he contended that the chirotheres had been large, semi-erect pseudosuchians, animals whose bones had not yet been discovered, who occasionally used their hands in running. They had claws on fingers and toes, as the tracks indicated. Their size could be inferred from the length of their steps, which meant that the smaller species had been four to five feet long, the largest as much as ten feet. 'Considering all their characteristics,' Soergel wound up his argument, 'the chirotheres lie within the boundaries of the pseudosuchians, without, however, actually corresponding to any specific family of pseudosuchians.'

Soergel sketched a model of the animal as he thought it must have been, and assigned these new pseudosuchians to a new family, the *Chirotheriidae*. Almost all palaeontologists applauded these deductions. Soergel had shown that Sherlock Holmesian methods could be used to arrive at the physical appearance of an

The tracks on the Hildburghausen sandstone slab which puzzled generations of scientists.

extinct animal from the tracks alone. Even Baron Nopcsa did not dispute Soergel. The baron had other fish to fry; disguised as a Rumanian shepherd he was once more back in the Balkans, on a political rather than a palaeontological mission. As the case of the chirothere entered its last stage, the baron was no longer able to take part in it. He had put an end to his hectic life by suicide.

The last chapter of the story of the mysterious chirothere (for the present) was written by Baron von Huene, who from 1927 to 1929 headed an expedition to southern Brazil to excavate saurians. One day a Brazilian named Vincentino Preso came to him and reported that he had found some unusually large bones in the vicinity of Chiniquà, in the state of Rio Grande do Sul. Baron von Huene found that the Chiniquà bones came from the Upper Triassic red marl. He had the block of stone in which the fossils lay embedded coated with plaster, wrapped in canvas, and packed in a crate; it was shipped to Tübingen for extraction and study of the bones.

In Tübingen it turned out that *Prestosuchus chiniquensis*, as the saurian was called after the names of the discoverer and the site, had been an imposing pseudosuchian almost seventeen feet long and more than four feet high. Moreover, it represented the very type of pseudosuchian whose existence Wolfgang Soergel had predicted. Soergel had every reason to feel triumphant:

374

down to the smallest details, *Prestosuchus* was the image of his chirothere model. And above all, the foot 'fitted' in appearance.

Nevertheless, there was a catch even here. *Prestosuchus* was a third longer than the largest conceivable chirothere reconstruction. The print of its foot, as Baron von Huene and Soergel at once realized, must have been twice as large as that of an average chirothere track. Thus, the animal resembled the imaginary chirothere but was undoubtedly not identical with it. Of course the Tübingen scientists had not really expected identity, especially since *Prestosuchus* had lived in a later epoch of the Triassic and in an entirely different region of the globe. The astonishing correspondence between the chirothere model and the Brazilian saurian was quite enough evidence that chirotheres had not been amphibians, dinosaurs, or any other ancient creatures, but must have been pseudosuchians. Moreover, discovery of *Prestosuchus* showed saurian experts that not all pseudosuchians had been small. They must have come in a great range of sizes – and therefore, by the laws of probability, also in chirothere sizes.

Nevertheless, the story of the Triassic chirothere to this day lacks its final point. No amount of searching has turned up a fossil really identical with a chirothere.

Jurassic: The Gingko Tree

When Darwin coined the phrase 'living fossil', he was not referring to any particular primordial animal species that had continued on across the ages, but to an unusually primitive tree
whose fanlike leaves deeply incised in the centre seemed to have
been dreamed up by a Biedermeier artist. That is the gingko tree.

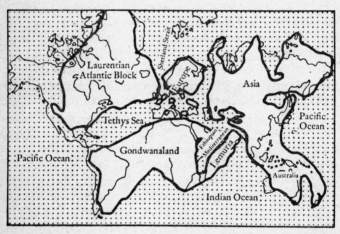

Distribution of continents and seas in the Upper Jurassic. *Based
on Bölsche, Frech, Koken, Arldt, Neumayr, and Laparent.*

Two hundred and fifty years ago the gingko tree existed only
in a few places in China and Japan. It was considered sacred and
was therefore cultivated particularly in temple groves. In one
such temple grove, near Nagasaki, Engelbert Kaempfer, a German physician and traveller, saw it in 1690 – the first European
to see, or at any rate to record seeing, a gingko tree. Several

decades later a few seedlings of the gingko arrived in Holland and England. When they attained some size, naturalists and plantlovers took such an interest in the tree's striking form and botanical peculiarities that they attempted to acclimatize it to western Europe. For two centuries a great deal of loving care was lavished on the strange Far Eastern tree in the parks and gardens of Europe and North America. Darwin admired it in the Kew Botanical Garden. Young specimens commanded spectacular prices. Today the 'living fossil' thrives in many regions from which it had vanished since the end of the Age of Reptiles.

Petrified Forests

Books on palaeontology usually have a good deal to say about fossil animals and very little about fossil plants, much to the distress of palaeobotanists. For both on land and sea the most significant event in the history of the earth has been the evolution of vegetation. Blue algae and other unicellular water plants created the basis for the existence of animal life in the seas and lakes. Plants were the first organisms to conquer the dry land, and made it habitable for the animals that followed them. Plants alone provided the animal world with food, shelter, and diverse environments, thereby establishing the preconditions for life and evolution. Changes in vegetation, the transformation of a swampy primeval jungle into an arid prairie, or a parklike savannah into a sun-drenched cactus desert, for example, caused the extinction of whole groups of animals and the rise of other groups. Even the conquest of the air would have been impossible without plants. An atmosphere conducive to animal life arose only after the land had already been occupied by relatively high types of plants.

Petrified plants have been found just as frequently as fossil animals. We have fossil algae, mosses, grasses, weeds, and flowers, fossil pollen, fossil trees, fossil resin, and even whole fossil forests. Possibly these plant fossils have been somewhat outshone by animal fossils because they do not differ so markedly from the vegetation of the present day. Archaic psilophytes, the scale trees of the Carboniferous, the palm ferns of the Mesozoic and other botanical fossils do not strike us as so startling and strange as many of the animal families that have vanished forever.

Our present mosses superficially resemble those of the Silurian period, in which naked, leafless, irregularly branched sporophytes were the first of all organisms to venture on to the land. If in our minds we enlarge our present-day ferns, horsetails, and club mosses to the size of trees, we can have a good picture of the swampy forests of the Carboniferous. The palm ferns, of which we still possess some sixty-five species in the tropics of South America, Africa, and Australia, began their careers in the Permian, developed with special luxuriance in the Mesozoic era, and have left behind highly interesting fossils, especially in the New World. Because of their strange form fossil palm fern remains are popularly called 'petrified beehives', 'petrified cactuses', or 'petrified pineapples'. The Black Hills of the Dakotas contain an unusually large fossil palm fern forest.

But a significant number of plants of the Jurassic period are with us to this day. One of these is Darwin's living fossil, the gingko tree. The bizarre araucarias are another, and can be found in South America, Australia, and Oceania. The mammoth sequoias of western America are not only the oldest living things on the planet, but also direct descendants of Jurassic trees. Relatives of theirs are found in New Zealand and the Far East. To be sure, the sequoias only reached their climax after the end of the Jurassic, in the Cretaceous and especially in the Tertiary. The famous petrified mammoth tree forests in Yellowstone Park date from the Miocene, the Middle Tertiary.

Among all these remarkable survivors of earlier ages, the most interesting is undoubtedly the gingko, for a number of reasons. It looks like a deciduous tree and bears two-lobed, fan-shaped leaves; but in reality it is closely related to the conifers. Deciduous trees did not appear until the late Cretaceous – and the fossil remains show amazingly 'modern' forms: poplar, beech, oak, willow, birch, laurel, and other such species, along with magnolias, water lilies, and other flowering plants. The deciduous trees differ from the conifers in bearing seeds enclosed in an ovary. Such trees did not yet exist in the Jurassic; 'naked-seeded' gymnosperms dominated the landscape.

Gingko trees are sexed; each tree is either male or female. The male bears catkins at blossoming time, resembling those of the hazel tree; the female tree bears stemmed flowers. Fertilization takes place as in sporophytes and palm ferns, by means of motile sperm cells. The yellow fruits hang down in clusters on short stems and look like a cross between apricots and cherries. The meat is edible, as are the kernels. All in all, the gingko represents a very old and highly important group of gymnosperms. As the rhynchocephalians are surviving saurians and the coelacanths surviving primeval fish, the gingkos are surviving primitive trees. There were gingkos in the Triassic, and they appear also in the Cretaceous and the Tertiary; there is a continuous record of them down to the present time. But in the Jurassic period they developed with particular luxuriance, like their animal contemporaries, the saurians. During the Jurassic, gingkos were spread over the entire earth, from Greenland to Australia and Tierra del Fuego, from Europe to the Rocky Mountains and Mongolia. Herbivorous giant saurians browsed on their foliage and devoured their odiferous but somewhat insipid fruits. The archaeopteryx and the pterodactyl fluttered through their branches. Jurassic fresh-water formations along the upper reaches of the Amur River have yielded great quantities of fossil male gingko catkins, along with yewlike female gingko blossoms. We can assume that vast gingko forests framed inland lakes, lagoons, and ponds, and catkins, flowers, and leaves were blown into the water as the leaves and catkins of willows are today.

The Age of Adventures

As early as the time of Cuvier, geologists had observed that the Mesozoic deposits characterized by ammonites and ichthyosaurs occurred with particular frequency in the Swiss Jura Mountains. Hence Alexandre Brongniart and Alexander von Humboldt called the period Jurassic. Later on it became evident that Jurassic formations were also well developed in France and South Germany.

The Jurassic and the succeeding Cretaceous period has been called 'a single great adventure'. The phrase does not refer only to the tremendous tectonic changes. In the Jurassic and Cretaceous periods the sea again attacked the continental masses.

The conglomeration of landed areas ceased. The northern and southern continents split into several parts. In many regions of the globe new seas and straits broke through the land. In consequence, marine animals once again burgeoned. The newly created marine basins abounded with mussels, sea lilies, squids, crabs, fishes, and reptiles. But the really great adventure of Jurassic time was the rush to dominance of the giant saurians.

We have already spoken at length of these colossal mountains of flesh that swam in the Jurassic seas or tramped, waddled, ran, or hopped through the Jurassic forests and swamps. During the Jurassic period their empire seemed so firmly established that the small primitive birds and the not much larger primitive mammals scarcely seem to have made any mark. But there was another group of animals which had also reached a climax in its evolution, and certainly did not seem on the point of departing from the earth. These animals were the ammonites.

The ammonites, those primordial cephalopods, which like snails carry a shell made largely of lime, appeared for the first time in the Devonian and vanished towards the end of the Cretaceous. In the Jurassic they were so common that today they belong among the most important index fossils of that period.

In contrast to the snail shell, the ammonite shell is, in the words of Bernhard Hauff, 'centrically rolled on a single plane. The soft part of the adult animal's body was to be found only in the anterior chamber of the shell, the living room.' The ammonite could regulate its propulsive force by a siphon in its body. Like a majority of squids, the ammonites had developed a kind of rocket motor.

Ammonites have served as models for evolutionary theorists. The Göttingen palaeontologist Brinkmann used the ammonite genus *Kosmoceras* which appears in a continuous series of forms to show the rapidity with which one species can develop into another. Otto Schindewolf of Tübingen considered the ammonites ideal objects for study, their evolutionary history constituting a sort of scientific table from which to read off the time-span needed for evolutionary changes and for adaptation to particular environments.

According to Schindewolf, the development of the ammonites at first took place in an almost explosive fashion. Then, in the

Mesozoic era, came a long period of quiet, gradual changes by a series of small steps. In the course of these changes the ammonites became more and more differentiated and specialized. An overproduction of forms was followed by a 'phase of over-specialization and eccentricity of forms'. Thus the ammonites, one of the most highly evolved invertebrate animals, went the way of the saurians when new environmental conditions placed new demands upon them. They were, as Schindewolf has put it, 'no longer able to effect such a revision of their structure. They died out leaving no descendants.'

Their plant contemporary, the gingko, on the other hand, flourishes today under the hand of man and has proved to be as viable as it was in the days of the giant saurians.

Cretaceous: The Great Turning Point

The strata which have long been supplying man with his white writing chalk were yielding many distinctive fossils as early as Cuvier's day. The Maastricht Chalk provided the mosasaur, the Chalk of southern England the iguanodon and many marine saurians, the Chalk of the Seine basin marsupials, and the Chalk of northern Germany innumerable conches, snails, and other marine animals. It was thus almost mandatory to name the most recent period of the Mesozoic era the Cretaceous (from the Latin *creta*, chalk).

As time went on, the great finds from the American Cretaceous were added to these earlier discoveries: gigantic land saurians, sea-serpent-like marine saurians, toothed birds. Chalk has built up curious and grotesque forms of mountains: the chalk cliffs of England, sections of the Alps, mighty rock formations in western North America, the mountains of Saxony and Bohemia called 'Saxon Switzerland', the Adelsbacher Valley in the Giant Mountains, the islands of Rügen, and many other mountain formations. Even in antiquity chalk was mined, cleaned of stones, washed, and used for writing, drawing, cleaning, and polishing, as a vehicle for paints, to neutralize acids, to remove spots, and even as a fertilizer.

But just what is chalk? 'Fine-grained, soft, easily staining limestone of white, yellowish, or light-grey colour, usually contaminated by clay, glauconite, and iron compounds,' the mineralogists and chemists define it. They add that lumps of flint are frequently found in layers of chalk. Prehistoric men, who made their tools out of flint, often chose to settle in the vicinity of chalk cliffs.

The limestone of which chalk consists is largely organic in origin. In Cretaceous time the dominance of water was even more striking than during the Jurassic. The ocean flooded parts of Europe and North America, separated Africa from South America, and Australia from southern Asia; temporarily it even divided the continents of North America, South America, and Africa. The great southern continent of Gondwanaland was finally broken up. Although the ocean eventually had to relinquish much of its conquered territory, for the duration of the Cretaceous the world was distinctly watery. Even the land was full of swamps, lagoons, and jungles of reeds.

Wherever sea water covered the land, the remains of marine organisms were deposited: fragmented coral reefs, bryozoa, bits of mollusc and crustacean shells, and above all those pretty, microscopically tiny lime shells of foraminifera, which had already formed enormous sediments in the Palaeozoic seas. Every piece of writing chalk (if it is genuine, and not, as so often nowadays, made of moulded plaster) is a compressed accumulation of countless microfossils.

Only after the Cretaceous deluge receded did the mountains rise which grew to enormous ranges in the Tertiary and today form so prominent a feature upon the face of the earth. Volcanoes once again spewed their lava. Ice ages seem to have occurred, at least in certain regions. Since the end of the Cretaceous saw a great mortality of species the whole world over, this period is frequently invoked by those writers who doubt the slow processes of change propounded by actualistic geology and continue to believe in violent and abrupt revolutions.

During the Cretaceous the belemnites and ammonites in their innumerable fantastic and seemingly purposeless forms vanished, after having dominated the seas. The even more overwhelming and strange-looking orders and classes of saurians likewise disappeared. The reptile-mouthed toothed birds died out. And as though Nature had pressed a switch at the end of Cretaceous time, hordes of completely different forms appeared: insects, bony fishes, birds, and above all mammals, of which several groups soon developed into monstrosities as the saurians before them had done.

This great mortality, this radical change in the fauna at the boundary between the Cretaceous and the Tertiary, does indeed seem to the present-day observer to have been a revolutionary upheaval. But the Cretaceous period was no brief, rapidly passing event; it lasted for some eighty million years – which is to say twenty million years more than the time that has elapsed from its end to the present. In so long an age, which moreover with its floods, climatic changes, and shifts of vegetation was unusually eventful, it was inevitable that there should also have been many events in the realm of organic life.

Moreover, according to the latest researches of Adolph Knopf, George Gaylord Simpson, and other American scientists, no really sharp dividing line existed between the Cretaceous and the Tertiary. The two periods passed so imperceptibly into one another that it is all but impossible to say when one ended and the other began. Knopf and Simpson are inclined to reckon the last ten or fifteen million years of the Cretaceous as belonging to the oldest epoch of the Tertiary, the Palaeocene.

Giant Bones in East Africa

The mightiest of all Cretaceous organisms, the great dinosaurs, did not live only in America. Two particularly enormous dinosaurs, *Brachiosaurus fraasi* and *Brachiosaurus brancai*, have been found in the Lower Cretaceous of East Africa. They appear to have been so gigantic that even the famous North American *Diplodocus*, if set up beside them, would look in the words of their discoverer, Edwin Hennig, 'almost like their chick'. In Mongolia and China, Roy Chapman Andrews and Walter Granger, Otto Zdansky and other fossil hunters have turned up such monsters as *Helopus*, *Mongolosaurus* and *Tienshanosaurus* from the Lower Cretaceous. Another species, the colossal *Titanosaurus*, survived almost until the beginning of the Tertiary. It was, moreover, a cosmopolitan, for remains of it have been brought to light in North and South America, India, the Malagasy Republic, Rumania, and Great Britain, on the Isle of Wight.

Palaeontologists were surprised when they discovered that almost the same reptiles were to be found in the East African borderline Jurassic and Cretaceous strata as in the North American formations of the same age. The first person to happen on the track of these animals was a German engineer named

Sattler. In 1907 he was employed in Tanganyika, then a German colony, and was searching for semi-precious stones in the vicinity of the port of Lindi, below the striking dome of Tendaguru Mountain. Suddenly he stumbled over a gigantic petrified bone protruding out of the ground. He rubbed his injured shin and then, conscientious fellow that he was, reported the find to the governor in Dar es Salaam. The governor did what conscientious governors do in such cases: he had a report of the incident prepared and filed away.

Shortly afterwards the head of the Stuttgart natural history collection, Eberhard Fraas, happened to visit Dar es Salaam, in search of altogether different kinds of fossils. Dropping in at the German colony's club, he heard the story of Sattler's injured shin. He asked for the file, and immediately set out for the Tendaguru area. What he saw there convinced him that one of the biggest if not the biggest saurian ever discovered could be dug up in Tanganyika.

Eberhard Fraas did not have the pleasure of unearthing the Tendaguru saurian. He contracted a severe tropical disease in East Africa, from which he died eight years later after a long spell of invalidism. Wilhelm Janensch, the custodian of the Berlin Museum of Natural History, succeeded him in Tendaguru. Janensch was accompanied by Edwin Hennig, then twenty-seven years old, who was later to become a member of the Tübingen team of palaeontologists. From 1909 to 1911 Janensch and Hennig disinterred a whole reptilian fauna in southern Tanganyika which closely matched the saurians that Marsh and Cope had discovered in the New World.

'The remains of our giants,' Hennig later related,

that is the bones of the legs, the vertebrae, and the teeth, were almost without exception strewn confusedly in the embedding rock. Whenever we thought we were gradually assembling the parts of an animal, some saurian goblin would play us a trick: three like thighbones, two pelvises, or something of the sort would almost always suggest the presence of other specimens of the same species on one spot – but for ever so long the skull, which we particularly wanted, would not turn up. ... Gradually, however, the legendary creatures beginning to arise from their graves became stranger and stranger; they kept us constantly in suspense. Thus several elements of legs appeared, and before we could properly free the joints they looked like strong femurs. After days of widening the shaft, a considerably larger piece would be added, and then it would turn out that what we had been dealing with was only the metacarpal bone. Quite a glove those hands would have worn! These bones would now be joined by the tibia. Then after a while the humerus belonging to the same animal would turn up.

With interruptions due to periodic rains, the excavation of this tremendous brachiosaur took no less than one and three-quarter years. But the long labour was well rewarded, for the animal had been completely preserved, including the desired skull. Only in Berlin, as Hennig tells the tale,

could the last delicate excavation be completed: complete cleansing from adhering soil, soaking of the bones, joining of the fragments, and so on. . . . In 1911 all the pieces had finally been gathered together in the Berlin Museum of Natural History. But for the preparation and mounting of this giant – which went on despite the First World War – no less than twenty-six years were needed! Not until 1937 was the astonishing creature stationed in the great hall of the museum, which was barely big enough for it. Its spinal column measured seventy-five feet in length, its ribs alone eight feet; the neck was over forty feet long.

In the meantime Edwin Hennig had long since returned to East Africa and brought home new finds. The brachiosaur resembled the corresponding giant dinosaurs of North America. A counterpart to the heavily armoured stegosaur of North America was likewise found in Tanganyika, the spiny-tailed saurian *Kentrurosaurus*. This African reptile possessed a second brain, the spinal ganglion being ten times greater than the cubic mass of the brain proper. Predatory dinosaurs and small bipedal dinosaurs were also found in Tanganyika. In fact the world of African Cretaceous reptiles corresponded in so many points with that of North America that palaeontologists found themselves faced with perplexing questions.

'How are we to conceive the paths of migration by which these giants diffused over the earth without excessive differentiation?' Hennig asks. 'We know enough about the geological history of the continents and oceans to say that at this late period (that is, in the Jurassic or Lower Cretaceous) no land connexion existed between Africa and North America.'

The giant dinosaurs cannot have twice developed into the same forms in different regions. Nor could they have swum across the oceans. Hennig postulates a possible land connexion by way of Antarctica; but because that continent is now covered by a mantle of ice, the hypothesis cannot be explored.

The Baroque Era of Zoology

The world-wide disappearance of the giant saurians seems as great a mystery as their world-wide dissemination. As Othenio Abel stresses, it is highly improbable 'that in all these different habitats different forces acting independently brought about the extinction of the titanic saurians'. Yet the same causes must have been operative everywhere, he argues – perhaps 'organizational conditions such as degeneration, circumstances which somehow affected reproduction and which affected all members of this family, no matter how enormous the distances between them, at approximately the same time'.

But some naturalists find nothing surprising in the extinction of the saurians. They argue that no special climatic conditions, degeneration, or pathological phenomena need be invoked to explain the doom of the giants. Saurians, after all, did not just vanish suddenly in the Cretaceous; some of them had already departed from the scene long before, in the Permian, the Triassic, the Jurassic. Sooner or later, these naturalists argue, every family of animals dies out and gives way to new forms.

'There has been much discussion about the causes of the rise and fall of the saurians,' Fritz Kahn writes.

But why? Are not two hundred million years enough for the dominance of a family, and a hundred million years for the despotism of giants? Is not Olympus vacant now and the Parthenon a ruin? Why do we expect any stock to be immortal? Everyone knows that everything mortal is mortal. . . . Families die out; so do nations, races; and so too have the saurians disappeared. It only seems rapid to us because it lies so far back in the past and because in a hundred million years, one can toss ten million around as if they were nothing.

Some tangible causes may have contributed to the extinction of one or another group of saurians: over-specialization, climatic change, the increasing competition of the rising mammals, volcanic eruptions with floods of lava and rains of pumice – all these factors have been suggested. Volcanic ash, one theory goes, may have darkened the sky for so long a period, and so reduced vegetation, that the giant saurians suffered from shortages of food. Periods of drought and ice ages could also have decimated the sensitive colossi.

A curious, somewhat comic hypothesis ascribes the death of the saurians to the disappearance of the extensive Mesozoic jungles of fern. 'Ferns contain purgative oils,' Kahn explains, 'and thus with the disappearance of the fern forests the giants lost their accustomed laxative. The great masters of the earth became depressed, indifferent about fighting and sex, and so they ingloriously died out from constipation.'

One of the expeditions to Mongolia and Central Asia which Henry Fairfield Osborn had organized on behalf of the New York Museum led to an astonishing discovery. Near Shabarakh Usu in the Gobi Desert the expedition's head, Roy Chapman Andrews, found eggs and even whole clutches of eggs that had been laid by dinosaurs. Sandstorms had buried both the saurians and their eggs. Some of the eggs had obviously been gnawed and sucked dry by ratlike primitive mammals. Thus the question arose whether these tiny mammals had not contributed to the extinction of the dinosaurs.

Nowadays a popular theory centres around supernova explosion – the sudden flare-up of stars. If our solar system had ever moved too close to such an event, the result would have been exposure to a tremendous increase in cosmic radiation. The concentration of such radiation might have been ten or a hundred times what it is today. Since cosmic radiation, as we have long known, can influence the genes, Soviet scientists have argued that the mass death of the saurians towards the end of the Cretaceous can be ascribed to a supernova explosion in the vicinity of the sun. While smaller creatures in their burrows were able to survive this dangerous radiation, the larger animals were exposed to the full force of it and died out. They have been sterilized by a star.

But interesting though it may be to wander in this maze of hypotheses and to exercise the imagination, we really do not need such conjectures. To quote the sceptical and witty naturalist Fritz Kahn once more: 'The saurians died as everything dies. . . . Grandeur comes and grandeur goes; and so, with the saurians, a grand creative era in the history of life on earth came to its end – the baroque era of zoology.'

The sudden and rapid development of the mammals which then began in the Tertiary has excited similar discussions in lay and professional circles. After the disappearance of the giant saurians the mammals occupied within a relatively short time and in a bewildering variety of forms all the places that the saurians had previously taken. After their competitors disappeared, they filled the vacuums promptly in 'phases of explosive evolution', as the palaeontologists put it. This mammalian explosion took place, of course, over spans of time which are 'short' by geological, not human, standards – that is, over millions of years.

The replacement of the reptiles by the mammals is the factor that prompted scientists more than a century and a half ago to assert that a new era had begun with the Tertiary. The name 'Tertiary' has long since been outmoded; the period is no longer regarded as the 'third' geological age. But since the Tertiary prepared the way for present-day geological conditions and present-day flora and fauna, palaeontologists are nevertheless still justified in considering that it marks the beginning of a new era in the history of the earth, the Cenozoic era.

Tertiary: Death in Amber

In his *Principles of Geology* Charles Lyell divided the Tertiary into an 'Early Recent Time', the Eocene, a 'Middle Recent Time', the Miocene, and a 'Late Recent Time', the Pliocene. Later it became evident that an 'Old Recent Time', the Palaeocene had preceded the Eocene, and that it was also necessary to introduce another epoch between Eocene and Miocene, a 'Less Recent Time', the Oligocene.

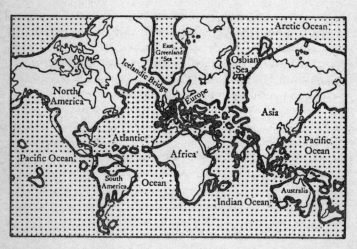

Distribution of the continents and seas in the Early Tertiary. *Based on Bölsche, Frech, Koken, Arldt, Neumayr, and Laparent.*

These five Tertiary epochs lasted, according to accepted views, about sixty million years. American scientists ascribe the last stage of the Cretaceous to the Palaeocene, thus giving the Tertiary a duration of from seventy to seventy-five million years.

As early as Darwin's time biologists discovered how the animal phyla of the present time developed during this period. Many of the forms that evolved were destined to vanish again. Others, however, proved viable; the future belonged to them.

The Tertiary was, after the Carboniferous, the period with the most luxuriant vegetation. Gingko, baobab, laurel, magnolia, cypress, and wine grapes flourished deep into the Arctic, while the Antarctic was host to ferns as well as a variety of deciduous trees. All the continents were covered with a green carpet, the carbonized remains of which subsequently became much of the world's supply of lignite.

The most common plant fossils come from an Early Tertiary conifer, *Pinus succinifera*, which looked much like a present-day pine. In the Palaeocene and at the beginning of the Eocene this tree formed great forests in the present-day North Sea region, and especially along the coast of East Prussia. The forests were a curious mixture of tropical and temperate-zone plants; for alongside *Pinus succinifera* and other conifers grew oaks, beeches, chestnuts, juniper shrubs, and also palms, palm ferns, cinnamon trees, and laurel bushes.

It is possible that the pines suffered from diseases. For their secretion of resin was unusually heavy. The quantities of fossil resin from the Palaeocene pine woods of East Prussia is estimated at between two and five million tons. Today they lie deep beneath the sand of the sea in a stratum called 'blue earth'. Year after year since prehistoric times the sea has washed small and large quantities of resin out of this layer and cast it on to the shore. For the coastal inhabitants this was always a joyful event. For this resin from a pine that had existed fifty to seventy million years before was already regarded, thousands of years ago, as an ornament almost as precious as the most precious stone. It is amber.

Amber Roads, Amber Monopoly, Amber Oath

Amber roads led from the Baltic and the North seas to the Mediterranean and the Atlantic. In Neolithic cultural strata, in Bronze Age urns, in pile dwellings, in barrows, in Mycenaean, Etruscan, Greek, and Roman ruins, amber jewellery has been found. The Phoenicians even carried the ancient amber trade to the Orient. All classical philosophers, historians, and naturalists were familiar with amber. They knew that it came chiefly from the region of the North Sea, where today it has become rarer, and they wondered about its origin.

There are many pretty fables about the origin of amber. In ancient Greece it was said that Zeus had killed Phaëthon, son of the sun god Helios, with a bolt of lightning for disobedience, and had transformed his mourning sisters, the Heliades, into trees. These trees thereafter shed tears periodically, and the tears hardened into amber. The Greek writers who told this tale certainly did not take it literally, but they already knew that amber is the hardened sap of a tree.

The Latin word *succus* means 'sap'; *Pinus succinifera* is the 'sap-bearing pine'. But it also means the 'amber-bearing pine', for *succinum* was the word for amber, as we may read in Pliny. Pliny even knew the kind of tree that produced the sap. When he set fire to a piece of amber he observed that it smelled like pine resin and burned like a chip of pinewood. Therefore it must be pine resin. Pliny reasoned that it had hardened 'by the influence of the sea'.

In the Middle Ages this knowledge was lost. It was assumed that amber – a favourite material for rosaries – was 'sea foam' mixed with salt and dried by air and sunlight. Some incipient geologists at the time of the Renaissance, including Agricola, held that amber was a mineral. Buffon and Linnaeus then demonstrated once again that the supposed mineral was in reality of vegetable origin. In 1757 Mikhail Lomonossov, the Russian chemist, asserted for the first time since Pliny that amber was resin from a tree. Ten years later Friedrich Samuel Bock, an East Prussian savant, conceived the notion that great pine forests must once have stood along the East Prussian coast and later sunk into the sea. In 1811 his fellow countrymen, Wrede and Schweiger, at last offered the explanation that amber was fossil resin from a long-past geological age.

So much for the natural history of amber. The fossil resin has played an enormous part in the history of civilization and economic life. For centuries the amber monopoly was in the hands of the Teutonic Order. When the knights of the Order were converted to Lutheranism, the Catholics stopped buying amber rosaries from them. Consequently Danzig merchants took over the monopoly and manufactured amber amulets for the Orient. In the time of the Great Elector the monopoly passed to the state of Prussia. To the inhabitants of the coast these monopolies were a great nuisance. They had to take 'amber oaths', raising their right hands and swearing never to gather a piece of amber along the beach. Anyone who did so and was nevertheless caught in the act by the monopoly's beachcombers would be hauled before an 'amber-court' and was liable to be hanged.

From 1811 to 1945 the collection of amber once again became a matter of private enterprise. The collectors fished the amber out of the sea, dug it out of the ground, and finally extracted it from the blue earth with huge dredges. In this way some six hundred tons of raw amber was mined annually in the amber centre of Palmnicken alone. Even so, the amber in the blue earth beneath the sandy ocean bottom has to this day, it is estimated, been exploited only to the extent of two per cent of the total available. Since 1948 amber has once more been made a state monopoly, for after the Second World War all of northern East Prussia became part of the Soviet Union.

In antiquity amber was fashioned into necklaces, buttons, rings, pendants, amulets, and all sorts of carved figures. In the Middle Ages and the Renaissance not only rosary beads but also frames, candleholders, bowls, small boxes, chests, powder flasks, portrait medallions, chess pieces, and household altars were made of it. In the seventeenth century even eyeglasses were made of polished amber. In modern times amber has served chiefly for the manufacture of cigar and pipe mouthpieces, varnish and lacquer. Since a piece of amber rubbed with a cloth will attract small bits of paper, the fossil resin has even given its name to a highly important phenomenon of physics. The Greeks called amber *elektron*; its attractive force is known nowadays as electricity.

In a Golden Tomb

Thus the amber of the Tertiary *Pinus succinifera* has become one of the most valued fossils to human beings, although it has passed somewhat out of fashion nowadays due to the competition of plastic resins. For palaeontologists it has in addition provided some of the chief evidence on small animal life fifty to fifty-five million years ago.

Insects and other such tiny animals are perishable organisms and are seldom found in the fossil state. But many of them were caught by amber resin and enclosed with it, so that they have been preserved in these shining golden tombs to the present day exactly as they were at the moment of death.

That gnats, ants, and other insects could be enclosed in amber was known to the ancient Greeks and Romans. Such pieces of amber were particularly sought after, and fetched high prices. The result was that in this realm, too, forgeries became common. Around the year 1500 skilful artisans devised a process for embedding small fishes, frogs, or lizards in a hollow in a fragment of amber. They then filled the cavity with linseed oil and sealed it with a patch of amber. Collectors paid huge sums for such curiosities.

Fish and frogs were never caught in the resinous flux of *Pinus succinifera*. To be sure, there is one authentic specimen of a young lizard so embedded. But usually the victims were insects or spiders. The imprisoned specimens indicate that the pine forests in the Palaeocene were inhabited by tropical and subtropical organisms. Termites, weaver ants, cicadas, praying mantises, sap beetles (Nitidulidae) and rove beetles are among the victims of the resin. What distinguishes these fossils from all others is that the organisms were suddenly caught and killed in the midst of life, in all possible situations. Thus the amber provides us with both a fine picture of the small fauna in the ancient pine woods and some notion of the mode of life and the associations of these organisms. Gnats on their nuptial flight and in copulation, predatory insects stinging and eating their prey, locust females laying eggs, ants migrating, butterflies emerging from the cocoon, spiders in their webs and with silk-wrapped flies, book scorpions clinging to the leg of an ichneumon fly in order to take a journey through the air – all such scenes have been found in amber and photographed. Some animals, it can be seen, resisted their fate with the utmost vigour. Daddy longlegs lost several of their legs in their struggles. Beetles whirled around, beat their wings, and tried so energetically to free themselves of the sticky liquid that they made deep furrows which are still visible in the amber.

An Amber Flea

Traces of vegetation are also found enclosed in amber: pollen, buds, whole blossoms, leaves, pine needles, bits of palm leaves, and crumbs of bark. Birds tried to pick out the insects stuck fast in the resin, and sometimes they lost a feather. The bird feathers in amber show that there were titmice, nuthatches, woodpeckers, and thrushes in the Tertiary pine forests, as well as sawbills, the motmots which today inhabit only the American tropics.

Even small mammals left hairs and sometimes whole clumps of hair in the resin. But it is virtually impossible to recognize from the hair what mammals these were. The imprints of small, four-toed feet in the resin point to the presence of predators, since the toes bore claws.

In one piece of amber Alfons Dampf, a noted entomologist, discovered a splendidly preserved flea. It was a species closely related to the fleas which today are parasitic on the Southeast Asian squirrel shrews called tupaias. The tupaias are genuine connecting links, transitional forms between insectivores and lemurs. The amber flea therefore seems to indicate that this genealogically important animal at one time lived in Early Tertiary East Prussia.

Amber even provides evidence of the presence of large mammals in the pine forests, for among the embedded insects are gadflies, those stinging, blood-sucking nuisances which torment animals and men on hot days. Female gadflies live on the blood of large warm-blooded animals, and must incontestably have done so in Early Tertiary times as well.

Whether the pines of the amber forests were actually diseased, and died from an unusually heavy outpouring of resin, is a question not yet settled. In New Zealand there is a relative of the *Araucaria*, *Dammara australis*, which secretes a white resin in such quantities that the branches and twigs sometimes look as if they are coated with icicles. Other species of *Dammara* produce so much sap that it can be tapped for industrial uses.

Perhaps *Pinus succinifera* was a tree of this same habit. But we do not know. The amber pines which have left behind so full a record of the past became extinct in the Eocene. What has remained of them, along with a few needles, seeds, and bits of bark, is their beautiful, shining golden resin.

Quaternary: Before Adam

At one time geologists and palaeontologists termed all post-Tertiary formations 'Quaternary'. Nowadays the Pleistocene, in other words the Ice Age (formerly called the Diluvian), and the Holocene, the present (formerly called the Alluvian), are reckoned as part of the Quaternary.

Geologically speaking, the Quaternary is only a very brief period, far shorter in duration than the other periods in the earth's history. According to accepted calculations, the Pleistocene and Holocene epochs together include only 620,000 years, of which only 20,000 are counted in the Holocene, the present epoch. If, however, the American scientists Ewing, Ericson, and Wollin are right in their new datings, based on deep-sea researches, the length of the Quaternary must be more than doubled. But even one and a half million years are a modest time-span compared with the epochs of the Tertiary; they are nothing compared to the length of earlier periods.

Like Tertiary, the term Quaternary is completely misleading. It dates back to the days in which the Palaeozoic era from the Precambrian to the Permian was regarded as the first era in geological history, and was given the name Primary; the Mesozoic era from the Triassic to the Cretaceous was then called Secondary. As new formations were recognized, a Tertiary and Quaternary had to be added. Since in those days geologists reckoned only in thousands of years, approximately the same duration was ascribed to each of these four hypothetical eras.

Today we reckon in figures of nine and ten places. If it is true that life is about a billion years old, then the Primary lasted about 840 million years, the Secondary 140 million years, and Tertiary 60 to 75 million years, and the Quaternary 620,000 to 1·5 million years. These divergent figures alone show that the old datings are thoroughly outmoded.

If we nevertheless continue to use the name Quaternary (and many geologists have discarded it), we do so out of our incurable tendency to regard geological and natural phenomena anthropo-centrically, to make ourselves the measure of all things. The Quaternary, which from the geological and biological point of view is only the last and smallest subdivision of the Tertiary, re-mains an independent period in the textbooks only because this was when our own species developed. This 'fourth age' is not a genuine natural subdivision of geological history, but is never-theless the age of man.

Yet even this statement calls for many reservations. Humanoid creatures already existed in the Tertiary. And since the palaeon-tologists, in investigating the fossil remains of the higher primates, often have great difficulty in determining where the animal ceases and man begins, there is no criterion by which we can establish where the age of non-men ended and the age of man began.

Modern investigators of prehuman times and human pre-history speak instead of an 'animal-man transition area' ex-tending from the Late Tertiary into the Pleistocene. The 'transition area' was preceded by a 'hominization phase', a gradual development of human traits; according to Heberer, it lasted for from four hundred to six thousand generations, that is, from twelve to eighteen million years, and began as far back as the Miocene, in the Middle Tertiary. If, accordingly, we wish to assign a special age of his own to man, we would have to make the beginning of this age not where the first Pleistocene glaciation began, but where the hominization phase started, or at the latest where the first representatives of the animal-man transitional area appeared on our planet.

But established terms, even if misleading and unfortunately chosen, are as difficult to root out as tough weeds. People still speak of the Diluvian, although we learned long ago that the epoch in question was actually an ice age. And people will go on speaking of the Quaternary and refusing to allow their own epoch to be reduced to an appendix of the Tertiary. Moreover, when we think of the tremendous changes that man has wrought upon the crust of our planet, we may concede that to predicate a special age of man has its deeper philosophical if not scientific justification.

A History of Primates

We human beings are counted among the primates, the 'chief animals'. It would seem that the primates, like other mammalian orders, evolved very rigorously in the Early Tertiary. Their ancestors were insectivores; and as we have mentioned, the connecting link between the insectivores and the most primitive primates is the tupaia whose fleas have been found in Palaeocene amber.

Small and insignificant though they look, tupaias are exciting creatures. They are highly prized study subjects for modern behaviourist psychologists. The palaeontologist Erich Thenius says of them:

They unite characteristics of the insectivores with those of the lemurs; in the past they were considered insectivores, whereas today they are regarded as the first representatives of the lemurs. ... They are protoprimates which have scarcely altered at all since the Early Tertiary, more than fifty million years ago, whereas the entire evolution of the primates and higher apes took place within that time-span.

Although lemurs and apes, which are mostly jungle dwellers, have been preserved in the fossil state far more rarely than animals of the plains, a great variety of primates from all epochs of the Tertiary period has nevertheless been found. Of particular importance to our own prehuman history are the Tertiary anthropoid apes, for some representatives among them took a step fraught with enormous potentialities – the step in the direction of humanity.

As late as the time of Darwin only a few fragmentary, puzzling remains of primitive anthropoid apes were known. An early ape with remarkably small teeth from the Lower Oligocene of Egypt, *Parapithecus fraasi*, and which lived more than forty million years ago, for a while attracted the eager attention of evolutionists, for it seemed to represent a transition to the higher apes. A gibbon-like early ape, *Propliopithecus*, likewise from Oligocene Egypt, was hailed even more enthusiastically. *Pliopithecus*, whose fossil remains were discovered in Miocene France, Switzerland, and Syria, appeared to be even further advanced along the road to man. Middle Tertiary gibbon-like apes later turned up in Asia and East Africa.

Today gibbons are restricted to southern Asia. Their world-wide distribution in the Middle Tertiary caused some biologists and palaeontologists, including Ernst Haeckel and Eugen Dubois, to assume a hypothetical proto-gibbon as the common ancestor of anthropoid apes and prehumanoids. But this theory did not stand up long because fossil representatives of actual anthropoid apes began coming to light.

One was a femur from the Eppelsheim sands near Mainz which looked so humanoid that, when it was first found, at the beginning of the nineteenth century, the philosopher Schleiermacher thought it the bone of a human child. There was a humerus, together with a few fragments of skull, dug up by Edouard Lartet in 1837 near Saint-Gaudens on the northern slope of the Pyrenees. There were also a few teeth from South Germany; they were found in Lower Pliocene strata near the villages of Salmendingen, Ebingen, Trochtelfingen and Melchingen in Württemberg. For a long time they were alternately regarded as the teeth of primitive stags, pigs or human beings.

After the theory of evolution began to dominate scientific thinking, all these finds were examined more closely. It turned out that the teeth and the bones from Saint-Gaudens and Eppelsheim belonged to a distinctly human-like ape, *Dryopithecus*, of which more remains were subsequently found in the Miocene and Lower Pliocene strata of Europe and Asia. *Dryopithecus* became a key figure in the theory of the descent of man. For a long time it was considered the animal ancestor of man, or at least, as Carus Sterne put it, a creature that had 'stood closer to the ancestral line of man than any living anthropoid ape'. The French palaeontologist Albert Gaudry briefly called it 'Tertiary forest man', and endowed *Dryopithecus* with the capacity to make tools out of stone; later he retracted this opinion.

Asia, too, had its Tertiary anthropoid apes, such as *Siva-pithecus*, an animal close to an orangutan, discovered by Richard Lydekker in the Sivalik Mountains of northern India towards the close of the last century. *Ramapithecus*, a far stronger ape, was unearthed from the same strata in 1935 by the American palaeontologist William King Gregory.

The Dryopithecines, as the whole group was called, were in fact fascinating creatures. They probably still lived in trees, but lacked the excessively long arms and shortened legs of present-day anthropoid apes. Consequently, they were not jungle acrobats, but unspecialized apes which could probably move quite ably along the ground as well as in the trees. Their teeth seemed so manlike that Gregory ventured to declare that they were animals precisely at or close to the fork in the road where the evolutionary paths of anthropoid apes and human beings parted.

Proconsul

In *Meet Your Ancestors* Roy Chapman Andrews avers that the ape crossed the Rubicon when he left the treetops for the ground, walking and running to hide in the underbrush when danger threatened, rather than taking refuge in the trees. As the ape's brain developed, he had the impulse to pick up stones and clubs and use them to defend himself against predators and to kill other animals for food. When he began to shape bits of rocks and pieces of wood into tools, he had already stepped on the lowest rung of the ladder that was to lead to humanity.

Opponents of the theory of human evolution have frequently made fun of such hypothetical reconstructions. But the evolutionists have been proved correct. Andrews wrote *Meet Your Ancestors* in 1945. Three years later British scientists proved by new excavations in East Africa that the crossing of the Rubicon had taken place much as Andrews postulated.

In 1923 fragmentary remains of a manlike ape of the Early Miocene had been discovered near Koru in western Kenya, fairly near Lake Victoria. In 1931 the most noted anthropologist and ethnologist in Africa, Louis S. B. Leakey, together with his wife, Mary, and a staff of assistants, began systematically combing the Kavirondo Gulf and Rusinga Island of Lake Victoria. By 1934 Leakey's team had turned up a number of jawbones, teeth, and bone fragments, all from the Lower Miocene. That is, the remains were between twenty and twenty-five million years old.

On the basis of these finds one of Leakey's associates, A. T. Hopwood, described three types of Tertiary manlike apes in East Africa. One species, *Limnopithecus*, belonged to the group of pro-gibbons, the other, originally named *Xenopithecus*, was scarcely different from the Indian *Sivapithecus* discovered by Lydekker. The third species looked like a primitive ancestor of the chimpanzee; and since a popular circus chimpanzee named Consul was then winning plaudits in London, Hopwood called the Rusinga ape *Proconsul africanus* – freely translatable as 'African ancestor of the chimpanzee Consul'.

In 1942 a complete jawbone of *Proconsul* was found, and it became clear that the animal was less like a chimpanzee than Hopwood had assumed. In some respects it seemed more primitive than any anthropoid ape; but in other respects, in the form of the teeth, the arch of the jawbone, and the con-nexion between upper and lower jaws, it surprisingly resembled man. Leakey therefore determined to explore the strata of Rusinga Island until he succeeded in finding a complete skull of *Proconsul*.

Not Leakey himself but his wife, Mary, accomplished this feat. In 1948 she obtained a splendid *Proconsul* skull. Insured for five thousand pounds, the skull travelled on her lap throughout the flight to England, where it was examined by a leading prima-tologist, Wilfred E. Le Gros Clark, at Oxford.

During the following years the Leakeys and their assistants dug on Rusinga Island, at Kavirondo Gulf, on Lake Rudolf and in the mountains of eastern Uganda. Along with fossil molluscs, fishes, crocodiles, birds and predators, they turned up many more remains of manlike apes. These bones came from several hundred different specimens. Three species of *Proconsul* alone were brought to light: *Proconsul africanus*, the size of a chim-panzee; the somewhat larger *Proconsul nyanzael* and the huge *Proconsul major*, which had reached the size of an adult male gorilla. In addition there were several types of *Sivapithecus* and of the gibbon-like *Limnopithecus*. Thus a great many Miocene anthropoid apes had turned up in a relatively small area. Of them all, the various *Proconsuls* were of the greatest interest for the evolution of man.

The concentration of anthropoid apes in this area must have been enormous. Rusinga Island is only about nine miles long and five miles wide. Evidently, anthropoid apes were evolving very rapidly during the Miocene, and were present in huge numbers. The sudden appearance of a host of new species and genera is that phenomenon we have earlier called 'explosive evolution'.

The explosive evolution in the area produced not only the ancestors of present-day anthropoid apes, but of virtually all the creatures that were moving towards the transition between animal and man. The species of *Proconsul* belongs to this category. They did not have the frontal arch, the acrobat's arms, the prehensile feet and the strong canine teeth of the anthropoid apes; and they may have been able to descend from the trees to the ground and race swiftly across the plains from one grove of trees to another. 'The erect gait,' Heberer has said, 'might have been acquired in a landscape of steppe interspersed with trees.'

The Secret of Monte Bamboli

The unravelling of the early history of man is a story in itself.* Here we can mention only a very few of the milestones along the way from animal to man. As early as 1872 Paul Gervais, the French palaeontologist, found remains of the lower jaw in a lignite pit in the province of Grosseto in Italy. He decided that it represented a Late Tertiary ape akin to the long-tailed monkey. The find was dubbed *Oreopithecus bambolii*, 'mountain ape from Monte Bamboli'. A single scientist disagreed with Gervais: Forsyth Major, who had found a fauna similar to that of Pikermi on the Greek island of Samos. Forsyth Major thought *Oreopithecus* must have been amazingly manlike. But no one gave him any credence.

* See Herbert Wendt, *In Search of Adam*, Houghton Mifflin Company, Boston, 1955.

In 1949 Johannes Hürzeler, the Basel palaeontologist, looked over the fragments of *Oreopithecus* in Paris and Florence. He concluded that Forsyth Major had been right. The teeth resembled those of an anthropoid ape or a predecessor of man. At this point the German-American anthropologist Helmut de Terra suggested that Hürzeler undertake new large-scale excavations at Monte Bamboli. De Terra even obtained the necessary funds from the Wenner-Gren Foundation and Swiss philanthropists.

But Hürzeler encountered enormous difficulties. The management of the Baccinello lignite mine where he began his researches was far more interested in mining than in saving fossils. It is estimated that merely in the first fifty years of the twentieth century more than thirty skeletons of *Oreopithecus* had been found and destroyed again due to the mining operations, or lost because of carelessness and indifference. The mine should have been bought for science to prevent such losses. But neither de Terra nor Hürzeler could raise that kind of money.

To make matters worse, the lignite mining operations came to a halt between 1954 and 1956, and again after 1960, because of their unprofitability. Thus Hürzeler had only a short time in which to dig. He ascertained, however, that *Oreopithecus* must have lived about ten million years ago. On the basis of the few available remains he even ventured to state in 1954 that *Oreopithecus* had not been an ape, but a prehuman.

In the summer of 1958, when Hürzeler was on the point of closing down the project, his men discovered another skeleton of *Oreopithecus* in the roof of a shaft at a depth of more than 2,500 feet. Since the pressure on the roof at this point was extremely strong, the precious bones were secured only with great difficulty. But the real trouble came after Hürzeler had finally brought *Oreopithecus* into the light of day. For the Italian authorities, who had hitherto done virtually nothing to further the excavations, suddenly refused to release the skeleton, in spite of previous agreements. They prevented Hürzeler from taking it to Basel to study. Their change of heart may be laid to the worldwide press reports that Hürzeler had found 'a man ten million years old'. Although no one in Italy had troubled about the *Oreopithecus* skeletons that had previously been destroyed or lost, the Monte Bamboli fossil had suddenly become a rarity which the Italian government was determined to keep in the country.

In the interval the sensation has subsided. *Oreopithecus*, the scientists decided, was neither human nor prehuman, but an animal, an anthropoid ape. Nevertheless, it already had one foot in the 'animal-man transition area', as it were. 'Certain specializations,' Heberer writes,

characterize it as representative of a side branch. . . . It was certainly no direct ancestor of ours, but in all probability a subhuman anthropoid ape, rather exemplary evidence for an early form of hominids. . . . It proves that the hominid stock was already independent ten million years ago; it shows us also, however, that the hominids did not descend from long-armed arboreal apes of the tropical rain forests. For at that time, ten million years ago, in all probability there were not yet any long-armed anthropoid apes.

Nutcrackers and Skilled Prehumans

The Leakeys, with the help of their by now grown children, once more threw light on the next chapter in human evolution. Since 1931 the Leakeys had had their eye on Olduwai Ravine, an impressive erosion gorge in East Africa. In the lowest stratum of Olduwai they found a great quantity of extremely primitive stone tools chipped from stream-bed pebbles. Since this stratum had to be one and a half to two million years old, the Leakeys were convinced that the maker of these pebble tools was the oldest then known manlike being.

The Leakeys spent three decades exploring the deepest geological layer of Olduwai. They found, along with more pebble tools, a number of Late Tertiary animal remains, including a pig the size of a rhinoceros and a gigantic ostrich some thirteen feet high. Meanwhile, however, other finds in South Africa came to public attention. From 1924 on a great many manlike fossils were discovered in Transvaal and on the border of Bechuanaland. At first these were considered erect anthropoid apes; but Robert Broom, specialist in the problem of missing links, demonstrated that they were prehominids.

These South African fossils fulfilled all the conditions required for ancestors of the human race. They still had apelike skulls, but their teeth were distinctly manlike. They walked erect, made simple clubs out of the thighbones of hoofed animals, hunted a great variety of beasts, and may even have been acquainted with fire and the human practice of killing other human beings. During the first years after their discovery scientists described, with somewhat excessive zeal, six different types; by now these have been reduced to one species with nut-cracker teeth and a gorilla-like bony ridge on its skull, called *Paranthropus*; and a far smaller and more manlike species which was first called *Australopithecus* and then *Australanthropus*.

The South African prehominids seemed to constitute ideal human ancestors. But there was one difficulty. They had lived only on the threshold of the Pleistocene and in the Pleistocene, that is five hundred to seven hundred and fifty thousand years ago. But since Java, Peking, and Heidelberg men, the first genuine early men, appeared between three and four hundred thousand years ago, the prehominids were simply 'too young' to have been human ancestors. The interval between them and early man seemed too short to most biologists. We have by now some approximate idea of the time-scale of evolution; we know that in barely a hundred or two hundred thousand years an animal-like, ape-headed prehominid could scarcely develop into a tool-using, intelligent early man.

For this reason many anthropologists argued that the entire group of South African prehominids must simply have been a side branch which missed evolving into man. Other scientists, probably more correctly, theorized that South Africa may have been a last area of retreat for the prehominids. They hoped that sooner or later they would find much older prehominids who could properly be called the real ancestors of man.

The Leakeys took this latter view. They, too, felt that the South African fossils were too young, and the South African sites too remote. *Australanthropus* and *Paranthropus*, they decided, were only prehuman extremes that had survived into the Pleistocene as 'living human fossils'. Persistently, the Leakey family team continued to search through the lowest stratum of Olduwai, using small tools and even toothpicks, the splinters of bone they found being often too tiny for any other implements.

In the summer of 1959, just before the hundredth anniversary of the publication of Darwin's *Origin of Species*, the Leakeys succeeded in putting together out of 450 bone fragments a pre-hominid skull. In the next few years, more remains of this East African prehominid were found. With the aid of a new chemo-physical test, the potassium-argon method, the Leakeys found that the Olduwai prehominid had lived one and three-quarter million years ago. Subsequent tests made him somewhat older, from 1·8 to 2 million years old. *Zinjanthropus boisei*, as Leakey named the find, was therefore more than three times as old as the South African prehominids. He seemed to have been the maker of the pebble tools, and corresponded in every respect, in age as well as geologically, biologically, and culturally, to the requirements that evolutionists had postulated for the ancestor of the human race.

Only the huge nutcracker teeth and the gorilla-like ridge on the head, the sagittal crest, did not quite fit the bill. In these two respects *Zinjanthropus* resembled the excessively odd-looking *Paranthropus* type of South Africa. But soon the Leakeys were able to provide another ancestor from Olduwai. From 1961 to 1964 they dug up remains of the skull, jaw, hand, and feet of five prehominids which were far smaller and in structure of teeth, skull and limbs far more manlike. Their age is estimated at about two million years. Alongside the huge, ape-headed 'nut-crackers', therefore, Africa has also produced small and amazingly manlike prehominids who perhaps already possessed considerable intelligence.

At first the Leakeys provisionally named these 'skilled dwarfs' *Prezinjanthropus*; by and by they wanted to include them in our own species as *Homo habilis*. But closer examination showed that the dwarfs of Olduwai did not differ in brain capacity from the range of variation of other prehominids. Just as the Leakeys' *Zinjanthropus* resembles the South African *Paranthropus*, their 'skilled man' resembles the South African *Australanthropus*.

Probably the 'nutcrackers' migrated as far as southern Asia; some remains of the bones of Asian giants suggest this possibility. But on the threshold of the Pleistocene these huge prehominids, which probably lived a peaceable, vegetarian existence, died out as the dinosaurs and mammoths had before them. The smaller, more intelligent, and less specialized prehominids invented tools, built walls, hunted their animal – and human – contemporaries, and developed into mentally alert early men who gradually, in the course of the Quaternary, were to conquer the earth.

(Editor's note: In January 1967, at a news conference in Kenya, Dr Leakey displayed plaster casts of incisor and canine teeth and bits of the upper and lower jaw of what he termed the oldest known ancestors of man, Kenyapithecus africanus *and* Kenyapithecus wickeri *– evidence that now extends the history of the family of man to more than nineteen million years.)*

Before the Deluge

The descendants of prehominids created civilizations, waged wars, developed ethical doctrines, denied their past and to this day are fond of emphasizing their exceptional place in the kingdom of living organisms. They have eaten of the tree of knowledge; they have discovered evil as well as good. They became Prometheus and Cain, Faust and Herostratus. They lost their animal innocence; they let themselves be led astray, and will never be able to exorcize the spirits they have called up.

Only falteringly and reluctantly have they begun to reflect upon all that happened before, and to recognize that their own story began only in the last minute of the world's history. And when they happened on evidence that made it questionable that a wholly new world had dawned with their appearance on the scene, their pride rebelled. 'Be calm, it was only an idea!' says Thales in Goethe's *Faust*.

Today we know that the prehuman history of the earth and of life was not just 'an idea'. It is truly a sublime thought, Darwin wrote half a century after Goethe, that the Creator breathed the germ of all life with its varied possibilities into only a few forms, or perhaps even into only a single form, that the regular succession of the generations was never interrupted, that no general deluge ever destroyed the world, and that from such simple beginnings the vast succession of the most beautiful and wonderful forms has evolved and is still evolving.

The Deluge, that local event that took place perhaps five thousand years ago, in the Sumerians' Mesopotamia, became the symbol of human sin and punishment. Whether it actually took place, and for what reasons it has loomed so large in civilized man's legends and traditions, is a problem for archaeologists and historians. It has nothing to do with the history of the earth. That was written and preserved in stone before ever the Deluge came.

Index

change theory, 147–9, 151
Ewing, Maurice, 153, 397
Extinction phenomenon, 173–80,
198–201, 289–91, 387–9; over-
specialization and, 227–8, 248–9,
297–8, 381, 387

Family trees, 189–218, 243ff., 252ff.,
305ff. See also Darwin (Charles)
and Darwinism; Evolution; specific species
Ferdinand II de Medici, 16, 17
Ferns, 355, 356, 358, 378, 388, 391
'Figure stones', 23–4, 27, 45, 53–9ff.
Finches, 193–4
Finlay, George, 246
Fish, 20–25, 30, 44ff., 63ff., 78, 79,
380 (see also Marine fossils; Sea
shell fossils); ancestors, 245;
armoured, 245, 337, 342–6; bony,
342–7, 383; Devonian, 347–52;
evolution, 90–91, 95, 110, 119,
127–30, 214–15, 220–24, 231, 245,
347–52; Silurian ostracoderms,
337, 342–6; tracks, 363
Fish-dragons, 222–3. See also Ich-
thyosaurs
Fitzroy, Robert, 190–91, 195, 266
Fleas, 395, 400
Flies, 395; tsetse, 289–91
Flight, types of, shown, 235
Fluorine dating, 325–6
Flying dragons, 122–4, 229–36
Font-de-Gaume cave, 164, 167
Footprints, 363–75
Foraminifera, 329–31, 383
Forests, 349, 352, 354–7, 358, 377–9,
391–6
Fort Laramie, Wyoming, 308
Fort Wallace, Kansas, 274, 292
Foulke, William Parker, 269
'Fox ponies', 253
Fraas, Oskar, 220–22
Fracastoro, Girolamo, 18, 24
France, Anatole, 28
France, 72, 77–81, 93–106, 108, 120,
124, 125, 250; cave explorations,
126, 167, 176; Ice Age fauna, 176;
Jurassic, 379; Miocene primates,
401; Triassic tracks, 365, 368
Franconia, 62, 70, 119, 122, 229
Frankfurt, Germany, 242

Franklin, Benjamin, 264
Frederick III, Emperor, 32
Frederick William IV, 271
Freiberg, 82; Mining Academy, 83
Freicine, and saurian skull, 117
Frogs, 394

Gadflies, 395
Gailenreuth, cave of, 70–73
Galápagos Islands, 192–4
Galileo Galilei, 16, 17
Gallows Hill, 43, 44
Gamow, George, 11, 181, 182, 184
Garriga, José, 197
Gaudry, Albert, 249, 250, 369, 401
Gazelles, 249
Geese, 97
Geneva, 133, 242; Lake of, 137
Geoffroy de Saint-Hilaire, Étienne,
102, 103, 107, 110–11, 114, 264;
and Cuvier, 127–8
'Geological clocks', 321–8
Geological Institute, St Petersburg,
160
Geological Society, German, 150
Geological Society of London, 87,
145, 335, 336
Geological Survey, British, 335, 336
Geological Survey, US, 301ff., 333
Geologic time scale. See Time, geo-
logic
Germany, 64–6, 82, 119–20, 153,
219–24, 382 (see also specific
locations); bird fossils, 233, 239–
40, 241–4; Ice Age, 134, 138, 140,
150, 151–2; Jurassic, 379; pri-
mates, 401; theory of evolution
and, 206; Triassic tracks, 364–8,
371–5
Gervais, Paul, 404
Gesner, Konrad, 27–8
Giant lizards, 115–17, 118, 119,
225–8
'Giants', 30, 31–7, 64, 78, 299. See
also Dinosaurs; Dragons; Rep-
tiles; Saurians
Giant sloths, 110, 197–202, 275, 364
Giant tortoises, 193, 275
Giard, Alfred, 250
Gibbons, 401, 403
Gigantostracans, 344
Gills, 347, 348, 350, 351

426

The Pursuit of The Millennium
Norman Cohn

A new, revised and enlarged edition.

The Pursuit of the Millennium tells how the desire of the medieval poor to improve the material conditions of their lives sometimes transformed by prophesies of a prodigious final struggle between the hosts of Christ and the hosts of Antichrist, out of which the world would emerge as a new paradise. In the earlier centuries the poor who were inspired by such prophecies would set off on people's crusades or on flagellant processions, and end by massacring the Jews. But gradually their hatred became concentrated on the clergy and the rich. The last chapters of the book show how at the close of the Middle Ages certain groups planned to exterminate the wealthy in preparation for the Second Coming, and then to impose an anarcho-communistic order on the whole world. In the Conclusion the author analyses the significance of these medieval phantasies and strivings from the standpoint of sociology and social psychology, and suggests how they relate to the revolutionary movements of our own times.

'A haunting and significant book.'

TIMES LITERARY SUPPLEMENT

'A piece of great originality and power ... it deserves study and emulation.'

Sir Isaiah Berlin

'Full of rich, fascinating, interesting scholarship ... What a field he covers!'

H. R. Trevor-Roper in the NEW STATESMAN

Homo Ludens
Johan Huizinga

Introduction by George Steiner

The classic study of culture as play by the author of *The Waning of the Middle Ages*.

'*Homo Ludens* is the most important work in the philosophy of history in our century. A writer with a sharp and powerful intelligence, helped by a gift of expression and exposition which is very rare, Huizinga assembles and interprets one of the most fundamental elements of human culture: the instinct for play. Reading this volume, one suddenly discovers how profoundly the achievements in law, science, poetry, war, philosophy, and in the arts, are nourished by the instinct of play.'

Roger Caillois, editor of DIOGENES

'The unrivalled historian of culture has in his work on Man at Play prepared the ground for the interpretations of history as developed by Toynbee, Sorokin, Alfred Weber, Gabriel Marcel, and even Camus.'

HUMANITAS

'Huizinga's essay on *homo ludens* is one of the few works informed about the problem of man.'

Martin Buber

Paladin